I0486309

EDUARD LERNER

LONGEVITY
and
HEALTH

A NEW EPOCH IN MEDICINE

**Alzheimer's, Parkinson's, Drug, Alcohol and
Smoking Addiction & Hundreds of Other
Diseases Can Be Treated**

DECEPTION AND THEFT PREVENTED IT

Amsterdam 2011

LONGEVITY and HEALTH
Eduard Lerner
e-mail: e.Lerner@wxs.nl

All rights reserved including rights of reproduction. No part of this book may be used or reproduced in any manner without written permission of the Publisher, except in the case of brief quotations embodied in critical articles and reviews.

Copyright © 2011 by E. Lerner
Paperback 978-1-4457-3306-7
Hardback 978-1-4467-3039-3

"Scholars who have achieved success do not have a pleasant life; they are envied, feared, but most often they are hated. They have known envy and hostility from their colleagues, and even from their friends."

Professor Joachim Volted, director of Higher Research School at Technological University, Eindhoven, Holland.

Contents

Preface

I have wanted to tell my story for a long time, but something always has stopped me. Doctors never have enough time, especially those who do research and make discoveries. I have finally taken up the pen to write about my life, the practice of medicine, research, and the events that shaped the course of my life. Everything that I recount comes from memory, images rewound like microfilm and transferred meticulously to words on paper. A part of this book refers to my personal life and one of my favorite hobbies: collecting stamps. Besides that, I revisit my job, my patients, my envious colleagues or jealous companies, my misgivings about doctors, both good and immoral, my musings on medicine, and most importantly, details of the research that has motivated my entire practice. A part of this book is focused on explaining my research on curing nervous diseases with the inventions called ETDDS and EAG, which can lead to a longer life, an emerging and fascinating field called Longevity.

Physicians well know about the "blood-brain barrier," which protects the brain from harmful constituents of the blood and thus improves the nervous system. Medicines taken by injection or tablet enter the bloodstream at doses far too low to cross this barrier into the brain. Therefore, it has historically been deemed impossible to treat brain disease via intravenous methods.

Around 1950, while perusing a litany of medical literature, I used a body of research claiming that no blood-brain barrier exists for the two or three millimeters between the upper nasal cavity and the cranium. From this, I began to question whether that small area could be used to deliver drugs to the human brain.

I worked on this problem for many years and finally developed a method that allows a large amount of a drug to be delivered into the brain through the nasal cavity. I called this method "enhanced transnasal drug delivery system" (ETDDS). I have received sixteen patents for the components of this process, presented my findings at international conferences, and written about my methodology.

The most notable advantage of my method lies in the possibility of more effectively administering the same drugs that are currently used in tablets or injections. Certain medicines that need to reach the brain arrive at sub-therapeutic levels, at best, due to the blood-brain barrier. With ETDDS, the same drugs could be delivered between ten and a hundred times more efficiently, directly to the brain. ETDDS allows these medicines to bypass the blood brain barrier via the nasal cavity and be introduced at much higher concentrations. My study shows a five- to thirteen-fold increase of medicine delivery to the brain.

In addition, I have new ideas that are all realistic and possibly even better than any of my current patents. They could be the basis to finish the whole ETDDS project within the next two years.

One of my new inventions is a series of special exercises called the "Biologically Pump Strengthening Intrabrain Blood Supply" to strengthen the blood supply to the brain; resulting in curing and preventing many brain illnesses. These exercises take only up to ten minutes a day and can be used, not only by ill people, but also by healthy people to prevent damage to the brain.

Hundreds of millions of elderly people suffering from vertigo, deafness, tinnitus, tremors, dysarthria, ataxia of the limbs, as well as other conditions will find that their symptoms will disappear after four to five days of using my method. More about this method will be described in chapter 13.

To give some background on the problem of brain diseases, I refer to the 2007 report from the World Health Organization (WHO[1]). Here I will focus on central nervous system disorders, and the biggest part of them includes brain diseases. According to this 2007 WHO report, brain disorders affect up to one billion people worldwide, and that number is set to increase as the population ages. In the United States alone, more than sixty million Americans—one in five—are affected by a brain disorder. Scientists know of more than one thousand disorders of the nervous system, having many different causes and coming in many different forms, including neurodegenerative diseases, mental illnesses, and dementias. Some of the more familiar are Alzheimer's disease, Parkinson's disease, stroke,

---

[1]. www.WHO.org; Worldwide problem of brain disorder, 2007

encephalitis, meningitis, arachnoid cysts, Huntington's disease, certain kinds of infections, multiple sclerosis, anxiety disorders, depression, schizophrenia, personality disorders, sense organ diseases, addictions (nicotine, drug, and alcohol addiction), eating disorders, learning disorders, sleep disorders, chronic pain, traumatic brain and spinal cord injury, and so on. A big part of the one billion brain disease patients have some sort of addiction, like heavy smoking, drug abuse, or alcoholism. This means that of the roughly seven billion people currently inhabiting the globe, one billion of these are sick, meaning that *one in seven* of us is sick. An estimated 6.8 million people die every year from brain disorders. Brain disorders do not discriminate; they affect people in all countries, irrespective of age, sex, education, or income. WHO claims that unless immediate action is taken globally, the burden of brain disorders is likely to become an even more serious threat to public health.

The 6.8 million people who died of brain disorders last year brings the total number of victims in the last ten years to more than sixty-five million, almost as many as the seventy million victims during the Second World War (WWII). In fact, the problem is so severe that one can even say that the rising incidence of brain diseases worldwide is tantamount to WorldWarIII.

I have devoted much of my time to the invention of electroautonomography (EAG), a process that allows diagnosis of illnesses in the nervous system and internal organs. I have patents for this invention as well and have also presented its methodology to conferences across the world.

Both ETDDS and EAG are meant to assist in the treatment and prevention of brain disease and to prolong the active and healthy lives of hundreds of millions of patients with brain diseases, most of which are currently untreatable. My methods have created a realistic new way of treating these conditions. In chapter 11 I described the importance of my inventions by showing the potential effects it could have on the hundreds of millions of patients worldwide that currently suffer from Parkinson's disease, Alzheimer's, strokes, brain cancer or any other brain related illness.

Of the seven million doctors who practice medicine worldwide, there are roughly 85,000 neurologists in 106 countries. There are 270,000 hospitals in the world. Central nervous system (CNS) diseases require 35 percent of the total investments from all pharmaceutical companies, a total of more than € 200 billion a year. Many millions of doctors, hospitals, institutions and other medical companies have tried to find new methods to diagnose and cure illnesses of people, especially for the nervous system that is more complicated than any other area in science. The problem is that the internal structure of the human body is very complicated and has many millions, even billions of different structures: cells – neurons, fibers, capillaries (tiny blood vessels) etc. Therefore it is extremely important and much more difficult to make new inventions in medicine compared to other areas in science.

To confirm this statement I will give some examples. It is known that a human brain weights around 1000 -1300 grams and has around 25 billion cells – neurons. One human organism has so many tiny vessels that if you put them in a row, it will be as long as 100.00 kilometers and together they have a surface of 6000 m². The lungs have 400 million alveolar cells. All this information shows a part of the difficulties that you will encounter when trying to invent new systems to diagnose and treat human illnesses. As a doctor, I have taken pleasure in being able to invent and develop a method that may eventually treat billions of ill people. The cumulative advantages of my methods will become central to treating and preventing diseases and will directly contribute to extending a healthy and active life. This will become a new horizon in the field called Longevity.

If we are able to prevent many dangerous diseases in the central nervous system, the body organs will stay healthier, because all body organs are regulated by the nervous system. It is known that the term 'Longevity' is used for the phenomenon that people can extend their lives above 100 and probably 120 or 150 years. But after the age of 80, many old people start suffering from different illnesses of the brain. Therefore we must not only strive for 'Longevity' but also making people healthier in their longer life. To reach such a goal I suggest my two inventions ETDDS and EAG can be used to prevent or treat many illnesses of the nervous system.

The title of my book is 'Longevity and Health - A New Epoch in Medicine'. From now on I will use the term 'Longehealthyvity' in the book but the reader must know that this means a long and healthy life.

The methods that I will describe in the book will start a new epoch in medicine and hopefully will extend the healthy lives of people up to or over 100 years.

In Western European countries I found not only that my colleagues resisted my EAG invention, but also that they actually concealed attempts to walk away with my research. As a Russian citizen, I found it difficult to fight businessmen, companies, and other scientists because of my imperfect English and the cultural differences, as I learned how to approach life in the West. In Russia I had already learned that people were not honest, and the West was no different. When I first moved to the West, I went to The Hague to apply for a patent. There, the lady in the patent office advised me to be very careful, as scientists from the West would immediately try to steal my ideas. I laughingly dismissed this, thinking such things could happen only in Russia. Within a few years, I had changed my view.

Earlier, in Moscow, with my assistants and laboratory, I had been able to perform many tests on animals for both my inventions—the enhanced transnasal drug delivery system and the electroautonomograph—with positive results. I was then able to apply the same techniques to many patients with different diseases with positive results and without any complications.

After I moved to the West, many interested companies and colleagues advised me to perform these experiments again before applying the technology to patients. I did so in two renowned laboratories in Germany (Covance, in Munster and Pharmacology and Toxicology, in Hamburg). The results of these experiments were positive, and so I was permitted to start treating patients. The official reports from these laboratories are presented in chapter 11 and 12.

In 1996 I was part of the world famous Geneva International Innovation Exhibition, also known as "The 24 Salon International Des Inventions". I was awarded the Gold and the Bronze medal amongst a highly competitive field with both my ETDDS and EAG inventions. Besides these recognitions the town of Geneva gave me a prize "The Silver Plate" and I was handed a golden pen by the mayor.

Also included in chapter 11 and 12[2] are financial calculations and detailed information that graduates at the University of Rotterdam made about the different diseases. These calculations have been

---

[2] p.p. 138 -140, 167 - 169

confirmed as realistic and objective by the Dutch Association of Inventors (NOVU), the same organization that presented me with the award for Dutch Top Innovation.

In the appendix a letter from the NOVU has been inserted that states: *'In medical history it is well known that such discoveries have a new epoch in medicine. At the beginning of last century a Dutch doctor – Dr. W. Einthoven – discovered and developed an EKG for which in 1924 he won the Nobel Prize. An English man – Mr. A. Fleming – discovered penicillin and an American – Dr. S. Waksman – discovered streptomycin, for which he also got a Nobel Prize. Using Dr. Lerner's methods, hundreds of millions of people diagnosed with brain related illnesses can be cured. Herewith these unique inventions are so important for practical medicine that nomination for a Nobel Prize is quite well possible.'*

The purpose of my story is not only to convey my medical knowledge, but also to describe the difficulties I have had in using my methods to further medicine. The biggest roadblocks have been the jealousy of fellow doctors and attempts of large companies to further their economic interests by taking my work. Many of these people who first showed great interest, even to the point of supporting me with written declarations of their positive professional opinion, later became envious once they realized the magnitude of my inventions. When I lived in Russia, I thought such problems existed only in that country, so I moved to the West in search of opportunity. Here, I found myself facing the same problems as in Russia—not only in the Netherlands, but also Belgium, Switzerland, England, Scotland, the United States, and other countries. My story is one of challenges that I have always tried to overcome.

The reader and even many doctors will have difficulties in understanding the importance and usefulness of my inventions. Of course pharmaceutical and other medical related multinationals, but even doctors, as well as regular businessmen and private investors will be against the development and production of my invention because there is a potential risk for these parties of losing vast amounts of profit. However it is important to highlight that not all medicine can be used via the ETDDS method. A large number of medicines have to be adjusted to be able to work with iontophoresis and others have to be reviewed completely.

You will find in the appendix of this book the original critiques and references of my experimental results from internationally renowned companies and scientists. Some of these scientists and companies were honest with me, but others wanted to take my ideas away from me.

I have to apologize for the structure of the book, which will be a bit unusual sometimes because of my superficial English. I ask the reader to primarily focus on the content of the book: the most important results of my lifetime of research, and the stories behind it. With new investors, I can finish my 3 inventions in a year or two and then introduce my inventions into medical practice. From the moment I started researching during my first year at University in 1946 up to now (2010) I can honestly say that I have never met a single person or company that was really willing and able to help in the development of the invention of a medicine that can treat millions of patients without having different, less noble, motives such as profit or personal gain.

For the last 13 years I cooperated with a big business man from Frankfurt Mains and now he lives in Switzerland. His name is Klaus Graf. If I look back on my partnership with Dr. Graf, I can see that he only financially supported my project the first 2 years which I will describe in chapter 11 and 12. When I received positive results from the experiments, Dr. Graf stopped financing the research. And it seems to me that he always tried to keep the benefits to himself. The complete story about the cooperation with Dr. Graf that turned out to be very difficult for me, I will describe in chapter 15.

Those who are interested in my project, I advise to read chapters 11 and 12 about my inventions and their potential both for the medical world and for business.

Chapter 1

My First Professional Steps

In July 1941, my family, including myself, my mother, father, sister, and grandmother, evacuated from the Ukraine, a previously sunny, prosperous place, amid the explosions of German bombs. I was 13 years old. We heard the shattering noises of houses collapsing in fire and the cries and screams of our fellow dispossessed. We rode the train to Kharkov, which seemed to be very far from the front line, and then to Sverdlovsk in the Urals.

As I reflect on my childhood memories, I see myself as a boy in the little village of Mezenka, near Sverdlovsk, crippled by zero-degree cold weather and scraping my hands on frosted lumps of soil as I searched for the few errant potatoes still left in the ground. Here I was digging potatoes as a refugee while imagining myself as a surgeon. That winter seemed tough even by Siberian standards, and we had nothing to shield us but the summer clothes we had worn as we fled.

I hardly attended school that year, pre-occupied with helping the farmers outside in the fields harvesting their crops. Nonetheless, with the arrival of summer, I received a certificate for completing the seventh grade. If you consider that I finished six grades at a Ukrainian school and spent the seventh grade in Russia in the frozen field, you can easily imagine the poor quality of my education.

At the age of fifteen, I made my first adult decision, to enroll at the construction trade school that had been moved from Kiev to Sverdlovsk. I intended to have food on the table, a roof over my head, and professional skills too. My new friends at school were fellow evacuees from Ukraine, and older than I was.

Even under the conditions of evacuation, life settled down little by little. My parents were working, and in the spring we planted a small vegetable patch. When I came to visit my parents on vacation, not only did they feed me, but they always somehow found food that I could bring back to school with me. In hungry times, these small pieces of bread were treasured.

At technical school, students were fed worse than sailors on the Battleship Potemkin: a broth of nettles and half a pound of under-baked, liquid bread. We were constantly freezing, the dorm barely warmer than the outside. In the first year, rather than going to class, we unloaded coal from railroad cars. It was hard labor, and we grew up fast.

Some of my colleagues developed enviable survival skills. A classmate whose father was a shoemaker made women's shoes and sold them at the market. He saw I was barely making it and offered me a job helping him. Unfortunately, the first sandals I made under his supervision fell apart when a woman tried them on at the market. Unsurprisingly, that was the end of my cobbler's career.

I did not become a house painter either, though I tried to master this skill while assisting a painter who lived in our dorm. He brought me and two other assistants to a huge room that we were supposed to plaster and paint. I was put in charge of the ceiling, high above the ground, a rough and exhausting job. Before too long, I became dizzy and fell off the ladder, luckily surviving without fractures. It seemed clear that fate did not intend me to become a shoemaker or a house painter—though I could not imagine then what it had in store for me.

In November 1943, Kiev was liberated and our school was ready to reconvene, but the city was in ruins and life was still brutally hard. We had no dorms, and we all lived wherever we could. For some time I stayed with my aunt, but she was too poor to keep me, so I moved to Chernovtsy, where my brother had settled after leaving the army. He had some type of important job and could take me in, but the year was a loss as far as my studies were concerned. My knowledge was not growing, despite being given a certificate for finishing the first semester of trade school. This certificate entitled me to skip to the tenth grade.

My highest grades from the second quarter were all Ds. The principal called my parents and told them that I had to go back to the eighth grade, but due to my age and experience, I managed to convince him to keep me in tenth grade for another quarter. Winter holidays arrived and gave me two weeks to show what I was made of. I buried myself in the eighth-, ninth-, and tenth-grade textbooks. I studied eighteen to twenty hours every day, barely leaving my desk for two straight weeks.

The third quarter began and saw me on a tight studying schedule. I would come home from school, eat a bit, and then hit the books. My teachers noted my diligence and helped me out with extra tutoring. Gradually, I began to understand more. The third quarter turned into an endless, torturous marathon, with the starting line in the eight grade and the finishing line in the tenth. In the five months before graduation, I covered four years of school.

Nonetheless, I finished the third quarter without any Ds and the fourth without any Cs, and by graduation I succeeded in earning all As. I think this accomplishment was one of the most difficult and important in my life.

That year the Ministry of Education introduced gold medals for school graduates with all As, for best academic achievement and best behavior. Six people in my class competed for gold medals, including me. I took the sixth place overall, and all six students' documents were forwarded to Kiev, where the Ministry would decide who was worthy of the prestigious award.

The month of June passed and July began to slip by, but Kiev was still silent. Entrance exams started, but no medals had been awarded and no certificates of graduation issued. Finally, on July 30, four medals arrived, but not one of them was for me.

It was too late to apply to the University of Kiev, so I decided to attend the local medical school, even though I had no real interest in medicine at that time and applied simply to prevent wasting another year of my life. About three days later, I began my entrance exams in literature and the Russian language and scored a C. That was simply bad luck, because I was given a question we had never covered in class. Although I scored As in physics and chemistry, this did not prove enough to get into the medical school. I was on the verge of tears and felt sorry for myself because of all the energy I had invested.

After a couple of days I received a phone call from Comrade Didek, the first secretary of our local Party Committee, who wished to meet me. Didek's son was my classmate, and his mother was the head of the parent committee. She must have told her husband about me, because suddenly I was being invited to the local Party Committee.

On the Friday of the appointment I arrived at the first secretary's office. He greeted me warmly.

"So you flunked, Eddie? How could you, an "A" student, fail to answer the question?"

I explained.

"I cannot promise you anything, but I will do what I can. Monday, stop by the school director's office. His name is Dimitri Lovla. He is an old Communist. I think he will figure it out."

I flew home as if I had wings. I could barely wait for Monday, and I rushed to school first thing that Monday morning. The waiting room was packed. Apparently, everyone who had failed to get in arrived together with any parents, relatives, or friends who were able to put in a good word. I leaped toward the secretary and said that I needed to see the director. She laughed. "Are you kidding?" she said. "People sign up to be seen days in advance, and you want to see the director immediately?"

"I'm sorry," I said, "but I need to speak with the director. I was told he would be able to see me."

"Who told you that?" she asked.

"The Party Committee."

"Who was it exactly, do you remember? There are many people working there."

"Didek," I said.

The secretary's face and voice suddenly changed.

"Well, why didn't you say that in the beginning? What is your last name? Lerner? Hold on a minute." She dashed straight into the director's office.

I followed her. The director was an elderly man with eyeglasses, dressed formally. He was finishing up a conversation with a couple of visitors and offered me a seat while I waited. After the visitors left, he turned to me. "So you want to be a doctor?" he asked.

"Yes, very much so."

"I was told you were a good student. So how could it be that you received a C?"

I repeated exactly what I had told Didek.

"Let's do this. Right now, make a request to me in writing." He handed me a piece of paper and a pen, and dictated the text. "Tomorrow morning come to school with your clothing, and be

4

prepared to dig potatoes for a week. When you return, you can start classes. Keep in mind, we are making an exception for you. Many were not accepted to the university, and they wanted it no less than you do."

This was a standard policy in the Soviet Union: all colleges and offices had to help at the time of harvest. When I got back to the city, my request had been approved. Unfortunately, while I was shuttling between offices and potato fields, classes had started. My classmates already knew the names of bones and muscles, had received their white robes and put in time at the anatomy lab, and in general they felt like old hands. I had barely learned to stand the odors of the anatomy lab or the sight of corpses, but I caught up quickly. Before long, I realized that medicine was my true calling.

In the first year, we studied the skeleton, including the bones in the skull. We were short of educational aids, but the friendly lab worker gave me an entire skull to study. I was supposed to break it up into individual bones but had to boil it first. At home, I placed it in a big saucepan of water and put it on the stove. It took several hours to boil the skull, so I left it on the stove while I rushed to my next lecture. Unexpectedly, my mother returned home early, smelled the gas, and wondered what kind of meat was cooking. She removed the lid, only to see a skull in the boiling water … and promptly fainted.

For another lesson, I had to study the nerves in the brain. According to instructions, every two hours I was required to take pieces of brain tissue from one solution and place them into another one. I didn't have time to sit all day and keep replacing the tissue, so I asked my mother for help. She gladly agreed, moving the pieces from one solution to another, every two hours. Had she only known that these pieces were brain tissue from a human body, she would surely have fainted again!

In my first year I became excited about research, first in microscopic research—called histology—and then physiology. Professor Sklarov, chairman of physiology, offered me a research assignment in that department for my second year. I had to prepare a frog, inject a solution into its heart, and log the results on a kymograph, an instrument for graphically recording physiological processes. While instructive, the experiment ended up being rather monotonous.

I got wind of interesting research being done by the chair of pathological physiology, Professor Ivan Fedorov. Because the study of pathological physiology could not start before the third year, he suggested that I might have a problem convincing Professor Sklarov to allow me to begin early. Luckily, my argument made sense: I wanted to begin the research early so that I could continue it into the third year and develop more substantial results. This convinced Sklarov, and I began working.

Young and energetic, Fedorov was the complete opposite of Sklarov. His experiments were conducted on cats and dogs. The core of his research was a reflexogenic (nerve-containing) area of the abdominal aorta (artery). I wondered whether there was a similar area in the upper aorta and if it was supported by sympathetic fibers—the finest branches of the autonomous nerve system.

My research went well. I actually managed to locate a reflexogenic area. In fact, this research became the basis of my scientific work on the nervous system and my first successful experiment with this region of the body, later written up in Professor Ivan Fedorov's 1949 book on Experimental Physiologists.

This project brought me tremendous satisfaction. Unfortunately, my happiness was also tainted with anxiety. Exams were near, including one on physiology, and it was clear that Sklarov was biased against me because I had pressed him to move me forward. I was the first student to go into the exam room.

The professor glared at me. "So, Mr. Pathologist Physiologist, please pick a question card."

I picked a card, and the question looked familiar. I felt confident and ready to answer.

"All right then, let's see what you know."

Sklarov ran me ragged for an hour. He asked me questions that did not appear in any institutional programs, textbooks, or manuals, until even he looked tired. He reflected for a couple of seconds, gave me a hard look, and then wrote "Excellent" in my grade book. "You have my compliments, Mr. Lerner. Good job."

Professor Fedorov eventually moved to another city, Lvov (in West Ukraine), where he became director of the Blood Transfusion Institute, and then to Kiev, to yet a higher position. Once he left, the new head of the department uprooted everything that had to do with Fedorov's work and prohibited me from performing my previous experiments. Undaunted, I paid a lab technician, and we set up a table with instruments on the roof. I arrived at nine each night and then proceeded to do all of my experiments on dogs, which I could not finish with Professor Fedorov.

Early in the third year, I stopped attending lectures. I only attended the practical classes, and not all of them, either. I preferred doing my own experiments or staying in the pathology/histology lab. In the second semester of the third year, we started on a new course, topographical anatomy and operational surgery, taught by Nicolai Novikov, an intelligent, well-educated, and talented teacher. It was not surprising that I joined his advanced group; I had always dreamed of becoming a surgeon.

Although Novikov was the chair of human anatomy, he was a lecturer-professor. Prior to the war, he had worked on his doctorate in Moscow, but a fellow student had wrongfully appropriated his findings. Unfortunately, I too, would eventually learn how widespread the practices of theft, treachery and betrayal were in a scientist's life.

Novikov assigned me a research topic that was important and vital. At the time of war, many of the wounded could not have blood transfusions or injections due to extensive skin burns or their state of shock. In these cases, it was necessary to perform injections into the bone, called interosseous injections. Although bone is not sensitive to pain, the injections still caused pain, for reasons unknown. I was set to the task of discovering why the pain occurred and how it could be prevented.

This research took a year and a half and lasted through the end of my fourth year. I began by setting up a small lab in the department of pathological anatomy, chaired by Professor Shinkerman. There, I would dye nerve tissues. The method of dying nerve tissue with silver nitrate was further developed in 1887 by a Spanish anatomist and histologist by the name of Santiago Ramon-y Cajal. His method was unknown in Chernovtsy, but I mastered it by studying printed sources. It was a remarkably difficult process because bone is impossible to cut into pieces small enough to study under a microscope. The bone therefore needed to be first decalcified using acid, which made the tissues resistant to dyeing. It was thus not only necessary to master the method used by the Spanish histologist, but to combine it with others, thus creating my own. I finally succeeded in creating an excellent preparation of bone marrow nerves, and the entire staff of the pathological anatomy chair lauded my accomplishment.

The second phase was to study the path of blood from its introduction into the bone marrow to its dissemination into the body's larger circulatory system. In order to carry out this experiment, I worked with the X-ray specialist at the local hospital. At seven in the morning, I would bring in a dog, put it to sleep with anesthetic drugs, inject it with contrasting fluid, and then allow the technician to take pictures. (I set up a vivarium, a place with live animals, in my house where I kept a couple of dogs for the process. My mother maintained the area, never having the slightest clue of what was going to happen to the animals.) I was forced to pay for the experiments out of my own pocket, but the X-rays were vital in studying the circulation of the substance. I conducted a whole series of these experiments, and Novikov was impressed with the serious attitude I held toward my work and the reliability of my results.

With that phase done, I could move on to the third and most important: the clinical trial. It was necessary to continue the research on human patients who could not receive intravenous injections. By then, I had mastered interosseous injections and was often invited to perform them—even as a third-year student! I also managed to develop a less painful method for the injections called metasternal, or out-of-chest, injection, and my process was presented at several symposiums.

I delivered a report on my work to Novikov's group for operational surgery. This was followed by a report at a scientific student conference, where all the professors were present. Some of them actually participated in the discussion and admitted that my work was at the level of a good master's thesis. This assessment was confirmed by an all-republic student conference, where my work won first place.

In my third year, I performed my first independent surgery, a multiple finger amputation. A man's hand had been smashed in an accident, and I was allowed to perform the surgery under a professor's supervision. The surgery ended very successfully both for the patient and me, because it resulted in a good reputation for me and recovery for him. In my fourth year, the assistant of the department, with whom I worked the night shift, gave me a chance to do an appendectomy.

My First Professional Steps

I was fortunate to have excellent professors, and I learned a great deal from them. I hope they too were happy to see my thirst for knowledge and my eagerness to take in everything they taught—in practical classes, in the operating room, and in the classroom. Novikov noted my diligence and aggressiveness and referred to me as the "safebreaker of science" because I was rarely daunted as I overcame obstacles in solving scientific problems.

After the year was over, Novikov invited me to his home to share a bottle of wine and some snacks. This type of thing was high praise from Novikov and an honor that no student had ever earned before. After a few drinks, the professor asked me what kind of doctor I hoped to become.

To my response he exclaimed, "Oh, a surgeon! Excellent! But let me give you some advice, Lerner, because you are an adult and I feel I can be open with you. Let's say you have been assigned your dream job in Voroshilovgrad, in the surgery department. What happens if your boss turns out to be a petty tyrant? You can kiss your dream good-bye, because he won't let you near the operating table. As a result you will be an errand boy, rather than a surgeon.

"Do not think I am trying to scare you or talk you out of your dream," he continued. "I am just advising you to choose a medical field where you will be the least dependent on the boss. You have a good, analytical mind and are striving toward clinical practice, so what about becoming a neurologist? That way, even if the head of your department is a bastard, he will not have as much power over you as if you were a surgeon. One does not need anyone's permission to read research literature, or to walk into a ward and take a look at a patient. You will be your own master. A huge field of clinical research will open to you, to apply your strength, knowledge and skills. Neurology is the most interesting field in medicine, and there are enough challenges to last you a lifetime. Just think about it."

I took Novikov's words very seriously, and went on to choose neurology as my field. I had not been interested in being a neurologist before, but once I decided to become one, I hit the books and didn't look back.

Chapter 2

Doctor's Plot

I received my medical degree cum laude from the college's new dean, Professor Mankovsky. Mankovsky had transferred from Kiev to replace Professor Lovla, who had been forced to resign for "evident failure to understand the Party's personnel policy." Mankovsky was a neurologist who knew of me and had attended a presentation of mine. Upon graduation, he asked me about my plans.

I told him I had been assigned to the Voroshilovgrad Region (in Eastern Ukraine) and that I wanted to become a neurologist. He offered to write a letter to Ukraine's chief neurologist requesting assistance. Armed with the letter, I arrived in Kiev. The chief neurologist addressed the letter with a similar request to Iosif Taytslin, the chief neurologist of the Voroshilovgrad Region. Everything depended on him, but he dashed my hopes.

"Little depends on me now," he said. "Your fate is in the hands of Ivanov, the head of the regional health board, to whom you must go with your assignment. He sends young doctors to the countryside. Other than that, I am afraid I cannot be of much help to you. I will try to speak with the head doctor of the hospital, however, because we do need a specialist."

The chief doctor pored over my degree, my character references, and the scientific article that I had published. Then he and Taytslin designed a hush-hush proposition. A lab worker was needed at the morgue—a job that no medical school graduate would ordinarily take. Nonetheless, the chief doctor signed off that "E. H. Lerner, the graduate of Chernovtsy Medical School," was offered this position and had agreed to take it.

Ivanov was not thrilled. He fumed as he read the hospital's request. On the other hand, he had to keep the books straight and had to fill a job that had gone vacant for three years. He turned to me and raised his eyebrows in distress.

"With a good diploma like this, are you really going to work for the morgue?"

"I am," I said.

"Do you know how difficult, dirty, and unpleasant that job is?"

"I do. It's a job like any other job."

"Are you going to run away if you cannot stand it?" he asked.

"No. I am obligated to work for the entire three years."

He tried to talk me out of it, but in vain. With a sigh, Ivanov gave me the letter stating that I would work as the morgue technician, with the word "morgue" underscored in bold hand. The chief doctor took the letter according to the request of the chief neurologist Dr Taytslin and promptly signed an order appointing me as resident in the neurology department. And with that signature, I became a neurologist.

Over the next few years, I lived at the hospital, eating and sleeping there, studying every chance I had, reading everything that I could and seeking out ways to practically apply my knowledge. Dr. Taytslin, the chief of neurology, whose medical talents seemed to come straight from God, was my most important teacher. I would later meet quite a few famous specialists, scientists, and professors, but none had his wisdom, deep knowledge, skill, and professional instinct. He had studied with famous clinicians like Professor Neiding from Odessa and Professor Eizefovich from Kharkov, both educated in the German school of neurology, which was considered in the early twentieth century to be the best in Europe.

I had the good fortune of spending two years working with Taytslin, and I spent every day learning from him, never getting tired of taking in his knowledge, and conscious of the fact that I would never have another professor like him. Like all brilliant people, he of course had his peculiarities. For example, often while doing the rounds, a nurse would come up and report to him: "Doctor, you have a phone call."

"One moment," he would say and then go on with his rounds. Fifteen minutes with one patient, fifteen with the other ... then move to the next ward, until another two hours had passed, at which point he would suddenly exclaim, "One minute, I have to take a call!" He would then rush to the telephone, grab the receiver, and answer, "I'm listening."

Doctor's Plot

He also lost his galoshes almost every day. He would climb aboard a streetcar, take them off and forget them on his way out.

One day, a coworker witnessed him walking to work lopsided, one foot on the road and the other one climbing on the sidewalk every other step. She walked up to ask why he was walking that way.

"I did not know that I was...." he said, embarrassed. "And here all along I was wondering why I was limping!"

So, he had those types of peculiarities. And yet still he was a remarkable person.

I worked a lot in the neurology department, taught physiology in the medical school, and at the same time managed the Pavlov Society Group, organized for professors and doctors of the regional hospital. In 1904, Pavlov won a Nobel Prize, the first Russian scientist to do so, for his classical work in digestive physiology. His international authority in physiology was unimpeachable; his works truly started a new era for physiology. The Party's Central Committee thus decreed that all doctors should be studying Pavlov's work. Study of Pavlov's theory of the higher nervous system was a paramount task for the government, overseen by State Security and Party officials.

On January 13, 1953, every newspaper in the Soviet Union published the item "Group of Saboteur Doctors Arrested," which asserted that security services had exposed a 'Jewish medical organization' that used "harmful medicine to reduce the lifespan of Soviet public figures." A campaign of anti-Semitism followed the discovery of this so-called plot against the government, during which many Jewish doctors were fired, arrested, or even executed. This all happened in Voroshilovgrad, where I was situated. Luckily, I was merely fired, not arrested or executed.

There followed a period of collective confusion, inaction, and anxious anticipation. No one knew how this oppressive atmosphere would dissipate or whether the arrests would continue. Finally, on April 4, 1953, the newspapers published a report from the Interior Ministry stating that "after examination, it was discovered that the arrested doctors were wrongfully accused without any lawful foundation." The imprisoned doctors were pardoned, though two of them, Kogan and Etinger, had already died in jail. Lidia Timashuk, who had exposed the so-called "murderers in white robes," had to

13

return her decorations, medals etc. The head doctor of the hospital was then ready to hire me back, but I could not forget the humiliation and fear that I lived through during that time. I did not want to remain in Voroshilovgrad, and I headed for Chernovtsy in the western part of Ukraine without a single farewell. This period has since been called the "Jewish Doctor's conspiracy."

In Chernovtsy, which I had not seen since my departure from medical school two years earlier, I began to lecture and work for free in a clinic for nervous illnesses under Professor Sergei Savenko. Though my colleagues remembered me only as a student and had never experienced my work as a doctor, after a couple of months they regarded me fully as a peer and began inviting me to give consultations. A unique triumvirate came about, consisting of Professor Savenko; his assistant, a knowledgeable doctor named Rusinova; and me.

The three of us discussed every difficult case and formulated well-defended opinions that were easily accepted by other doctors. The neurology department was located on the base of a big psychiatric hospital consisting of many buildings and departments, and many doctors attended our conferences. Some of them told me that the reason they attended was to hear my reports. This was the highest form of flattery for a young doctor.

Despite my success as a neurologist, I worked for two years for free, unable to get even a part-time position. Luckily, time flew. I reviewed my thesis about interosseous injections, then bound it and presented it to the department of neurology as a candidate's thesis, roughly equivalent to the Western Master's. The attempt was successful. Now I was one step away from the degree of Candidate of Science. This step turned out to be a huge one and took me from Chernovtsy to Moscow.

Chapter 3

The Move to Moscow

I arrived in Moscow on December 21, 1955. In my hands I carried a suitcase with all my belongings, including my thesis. It was bitter cold that winter, and I nearly froze on the long trip to my aunt's house outside Moscow.

The weather did not get warmer that whole first month, as I ran around to clinics and hospitals in search of a job. In order to have a job, a residence permit was necessary, and in order to obtain a permit one needed a job—a vicious circle. In the Soviet Union, one could not move without a permit, and a new law had been introduced prohibiting out-of-towners from settling in Moscow due to overpopulation. These restrictions made an already difficult task a hundred times harder.

I do not know what angel guided me toward Kuntsevo, a large, factory-filled, working-class town across the river from Moscow. I got lucky; a clinic there was expanding and adding a new building. They needed doctors, and the old building had vacant rooms fully suitable for living. The chief doctor, upon discovering that I had worked in a clinic for nerve diseases and had a thesis to boot, offered me both a job and a room. That woman knew how to get things done; she knew how to deal with police, the State, and medical authorities—a priceless quality for a chief doctor.

Thus, I began working in the new building and was forced to live in the basement of the old one, where six former offices were now occupied by six different families sharing a small space. It was too early to celebrate. However, the construction outfit that owned the building had promised it to its own workers, so we had to move in quickly. At night, the clinic's van drove up to the back. We quickly unloaded our beds, tables, chairs, and dishes and fully occupied the premises. The following morning, when the construction outfit discovered our illegal move, the furious supervisor placed guards by the entrance and warned that whoever left the premises could not return. The construction workers threatened to throw us out on the streets along with our belongings.

The head doctor contacted the chief of police, who sent over a police officer. We now were under two different guards: the construction people and the police. We were in a state of siege. Patients were brought to our offices, as were food supplies.

We lasted a month in that basement while the chief doctor negotiated with the director of the construction unit. When they saw we firmly intended to stay, they surrendered and removed the guards. They never issued permits for our floor space, however, so we still lived illegally.

In 1952, I married a student that was 1 class lower named Isabella Teodor, and one year later our daughter Inna was born. My wife and I were happy. She and our child were finally able to move in with me. Good or bad, we had a roof over our heads, four walls, water, light, and heat.

Now that I had a job and a place to live, I needed to get my degree. Colleagues advised me to turn to a local professor named Gregory Kassil. I had met him in 1949 when I came to Moscow for a consultation regarding my work with Novikov. I told Kassil that in Chernovtsy I had developed a method to deliver medicine into the brain through the nasal cavity using iontophoresis, a painless and not unpleasant method, using small electrodes that give very low-level impulses that transport the medicine into the brain. He was very curious and questioned me about the details.

In the fifties, unbeknownst to me, he published an article on how to deliver medicine into the brain with the method I had told him about. (This is my method called ETDDS, which I will further explain in chapter 11). Now, in 1952, Kassil said he could not recall my name. I did not take umbrage, for he had helped me a lot. He read my thesis, approved it, and recommended that I defend it at the Central Institute for Continuing Medical Studies (TsIUV). Kassil arranged my defense with Deputy Director, Academician N. Grastchenkov, a famous neurologist and an important government official.

My defense went well. I often visited Grastchenkov at his department. I came to know all of the colleagues but I continued to work in Kuntsevo. It was boring, but I could not quit because they owned the housing.

Luck helped. I had a second job at a clinic with paying patients (as opposed to the free care provided at a State clinic) and one of my patients was a man named Vishnyakov, deputy secretary of the Moscow regional Communist Party Committee. He was very pleased with his recovery and took an interest in my work and my life. He was aware of my problems and decided to help me.

Once, he called and told me that his boss, Ivan Kapitonov, Moscow Party regional secretary, wanted to see me about his health. He was suffering from back pain. After a couple of acupuncture sessions, using methods I had learned from Chinese doctors who came to Moscow in 1958–1959, his pain was gone.

Kapitonov turned out to be a grateful person. He told me that I should move to Moscow. For any ordinary human being, was arranging this would be an impossible task, but for the head secretary of the province, all it took was to pick up a phone, say a couple of words, and that was it!

Vishnyakov sent me to the boss of Glavmosstroy, the city construction authority. His secretary cast me an icy glance, but when she heard that Kapitonov himself was to telephone regarding me, she turned pale and hurried to report to her boss. He came rushing out of his office, clearing it of visitors. "Wait until I call you," he told them. "I am busy taking care of something important. Doctor, please come in!"

He recovered his breath and called the director of the construction unit that owned the building. While on the phone, he erupted into curses and screamed, "Stop this crap—we are not at a meeting! Dr. Lerner will come to see you right now for a housing voucher and a permit for an exchange. Don't go home until you finish everything and report it to me personally. Understand? That's better."

He then turned to tell me, "Doctor, right now you take my car and go to Kuntsevo, and they'll take care of you."

I got into his black Isabell car and headed to the construction unit. The director met me in person. His secretary was there too, though the working day was over.

"Doctor, why did you not tell me that you needed a voucher?" the director of the construction unit asked. "I would have done it myself, because it was us who gave you the living space. Why bother Comrade Kapitonov? We respect doctors. Please have a seat."

That is how I received the voucher and the permission for an exchange. Of course, this was not only senseless but illegal as well—a voucher and a permission to exchange my residence permit to Moscow. Then again, what was legal at that time in that country? Permission from the Moscow government could take many months, even up to a year

The most important thing was for me to receive permission for the exchange. Yet I would gladly forget the process itself, for it brought nothing but headaches, wracked nerves, and mind-boggling expenses. After a tremendous amount of effort, we finally were able to move to Moscow. Our new place of residence was not far from Vorobiev Hills, where a huge building of Moscow University was already looming. We had a twenty-meter-sized room in a two-bedroom apartment. For myself and my wife, who at that time was also working on her thesis, our own room in Moscow now appeared a successful stepping-stone. We were extremely lucky, because the following year Kapitonov was transferred to Ivanovo to lead the local Party Committee. The year was 1958.

Chapter 4

The Defense

The next period in my life was filled with events akin to a Shakespearean drama. The heat of passion and the amount of plotting around me and my work could serve as basis for a thrilling novel.

Nikolai Grastchenkov left the TsIUV to head a chair at First Medical School, Moscow's most prestigious medical school. He took his staff and equipment and left behind tables, chairs, and naked walls. Professor Maria Tsuker, a neurologist who was famous even before I was born, stayed behind as a provisional head of the chair. She found herself in a tough spot and didn't know where to start.

When someone mentioned my name to her, she offered me the lowest position in the hierarchy, as a senior lab technician. This entailed a lot of work for very little money. Of course, though you couldn't live on 550 rubles, you wouldn't die either. You just ended up living the miserly life of a beggar.

Professor Tsuker told me frankly that she was hiring me out of great need, and I should not expect any promotions. I was happy to accept even these serf-like conditions. I had a chance to squeeze myself into science sideways and do the work that I loved.

Yet motivation alone is not enough. You need equipment, too. Firstly, I needed to set up an EEG lab, something I'd never done before. I also had to fill in some schooling gaps at the Institute of Neurology with Professor Zhirmunskaya. My studies there lasted about two weeks.

Sparing no time and effort, I searched Moscow for equipment. I was lucky to find a French-made Alvar electroencephalograph. With my own two hands I built a laboratory in my bathroom. This wasn't an easy job, but before too long my first technological "child" (the EEG lab) was finally born. I took care of it for about ten years, and later on, I mastered cardiography, plethysmography (research on the hands), and other modern tools.

The lucrative research I carried out there became my doctorate. I had enough material for three doctorates, but I needed a fivefold factor of safety. Like a bridge that the tanks needed to cross, I needed a thesis that would be impossible to break or destroy.

By that time, I had worked myself up the ladder and the government asked me to be a consultant and to help them. I had also become a consulting neurologist at the famous Building 3 of Botkin Hospital, where the entire Soviet moneyed class was being treated. It was not far from one of the oldest secret airplane factories, called Labor Banner, for secrecy. Its longtime director, Pavel Voronin, not only knew everything about building planes but also took care of his workers, technicians, and engineers. He gave my laboratory a truly royal gift, a Japanese polygraph worth $25,000, which helped me make difficult diagnoses for very ill patients. The factory bought it for us under a special arrangement, and we did everything we could for its workers in return.

I became especially close to the talented inventor Yuriy Dalago, the director of one of the factory's labs. Together, we constructed a unique test chamber for experiments on dogs, cats, and rabbits, since Botkin Hospital had an animal center. In the chamber, I conducted experiments that induced artificial strokes in the animals.

Shortly before that, there was a competition to fill a vacant neurology chair position. Professor Tsuker expected to get the job, which had been promised to her after Grastshenkov left for First Medical. Yet suddenly there was a change in the institute's leadership: Ms. Lebedeva, a knowledgeable and decent person, was replaced by Kovrigina, previously health minister. Kovrigina found herself in disfavor after she granted free stays at the sanatorium for Malenkov and Kaganovich, who had been banished from the Party "for creating anti-Party groups." After holding the minister's chair, Kovrigina's next position as director of an institute for doctors' continuing education must have felt like exile, but nonetheless, it was the demotion she received at the hands of her successor, Academician B. V. Petrovsky.

Tsuker's hopes of heading the department fell apart the moment Kovrigina got her job. Although Tsuker still had hopes that she

The Defense

became the head of the neurologist department but instead Professor Chetverikov became head of the department and she was left there as a mere professor. To her, that was a huge blow.

I soon became a junior researcher and then rose to senior researcher. Chetverikov needed tangible results, and I was working in a promising direction, dealing with patients suffering from strokes. They were often in states of coma, characterized by loss of consciousness and disturbed blood circulation and breathing. We observed the patients' symptoms and treated them successfully. This research became the basis for my doctoral dissertation, *EEG during a Stroke*.

I struggled with the idea of creating a medicine that would relieve edema of the brain. It became clear to me that people who are suffering from many brain diseases, are dying not so much due to hemorrhages or softening of the brain but due to recurring edema of the brain. Together with a smart graduate student from Mendeleyev Chemistry Institute, I created medicine for the edema of the brain. At first, we tested it in vitro and then on rabbits, giving them doses that were five times greater than normal and trying to detect the negative effect of the medicine on other organs and tissues. When we discovered no side effects, we next carried out a series of successful experiments on cats. We performed craniotomy, inflicted an edema, and then injected a drug intravenously. The edema vanished within minutes.

With such reassuring results, I filed for a patent. The request underwent a primary inspection to confirm that this was indeed an original drug but would need to be tested in clinical trials. To test it on people, we needed permission from the pharmaceutical committee of the Health Ministry. I was afraid to send in my request, as I was sure that the committee officials would immediately take the credit for my drug. This very important, necessary work thus remained unfinished.

Jealousy is a horrible feeling. Ancient people used to say "it is possible to cure any animosity, except that which is born out of jealousy." I felt that others were jealous of me because I was working on my doctorate and on my new drug and because I led the department in researching strokes. Every day I came across the most difficult patients, and the associate professors who were required to consult never even set foot in my department, assuming that I could not complete the job by myself.

Longevity and Health

One night, a patient who had previously been improving suddenly died, and I could not understand the cause of her death. I thought it could have been a defect in her oxygen tube, so I called my friend Engineer Dalago to take a look. Instead of a technological defect, he discovered traces of a crime. Someone had broken into the metal closet where oxygen tanks were being held and disabled the valve that controlled the gas supply.

Others were jealous that Chetverikov sent me all of his graduate students. In a year or two, under my supervision, they would develop a thesis and defend it successfully. All of this required a tremendous amount of energy, so that I hardly had any left for my own thesis. Nonetheless, I finally completed it in 1963, and it became the first doctoral dissertation in our department in the last thirty years.

I defended my dissertation in the Institute of Physiology at the Academy of Medical Research. Once again, Professor Kassil gave me helpful advice: "You need an outstanding advisor whose opinion is respected, either during the defense or the approval process."

I found such a person. He was Mikhail Livanov, an excellent physiologist, one of the founders of the Soviet electroencephalography. In Russia, after graduating from medical school, a doctor goes into a certain specialized area and then produces a type of scientific research called a candidate's dissertation. This candidate's degree of medicine in Russia is the equivalent of a PhD in the West. In the Soviet Union, in Russia, there are two dissertations. The first is called the candidate's dissertation and is more voluminous than the Western one. After the defense of that thesis, the doctor becomes a Candidate of Science. Usually, a candidate of medical science becomes a doctor after finishing the medical university.

I made my candidate dissertation when I was a student, during my third and fourth year. This was highly unusual. Years later, after the age of forty or so, the final doctoral dissertation is completed, which explains the accumulation of all the research and work one has done throughout his or her career. I finished my doctor's dissertation at the age of thirty-two. This made me one of the youngest Russians to have completed a doctoral dissertation. The defense of the dissertation gives permission for a doctor to receive the title of doctor of science, equivalent to the title of a professor. A doctor gets the actual title of professor when taking a professorial position. Alternatively, one can

The Defense

become a professor if, on top of defending a doctorate, he or she also supervises at least three candidate theses. I prepared at least fourteen.

My dissertation consisted of two different parts: clinical (the research done with EEG and other methods) and experimental (rabbits and cats were inflicted with a stroke, and we observed the results). For a few months, I did experimental work under the guidance of Professor S. Narikashvili from Tbilisi Institute of Physiology, headed by Ivan Beritashvili, one of the oldest Soviet physiologists. Narikashvili agreed to be my opponent during the defense of my dissertation, but he also suggested academic Livanov for my advisor. He went even as far as writing Livanov a letter praising my results and findings.

I brought the letter to Livanov's office. He took his time in giving me an answer. "Before I can say anything I want you to do a report in my laboratory," he told me. This was an invitation to a *corrida*, with me as the bull.

On the designated day, I arrived at the lab with two of my graduate students, Lapitski and Bibilishvili. My report dealt with the experimental part of my dissertation and did not take much time, but answering the questions that come with the defense of my reesearch took an entire day. It was an avalanche of questions, over a hundred, and I answered every one.

"So, are there any more questions?" he asked. Livanov glanced at his coworkers like a hunter looks at his pack of dogs before setting them upon a fast rabbit. They were ready to tear me to pieces, but they couldn't find a spot to pick on.

"Congratulations," Livanov said—and I'm quoting verbatim. "This was a heroic effort."

Livanov's approval was the highest reward for me. When he agreed to be my advisor, those who knew him refused to believe it. Perhaps that day should be considered as the real moment that I defended my dissertation. Soon the official defense took place too, at the Institute of Physiology. It was successful, and I became the one of the youngest doctors in the field of medicine.

After the defense of the doctorate, it was necessary to assemble all the important papers and send them to the highest commission of certification that awarded the scientist his degree and title, in addition to giving him the corresponding document. Scientific work fascinated me so much that I didn't feel like wasting days and even weeks on collecting the necessary papers and documents. I didn't submit my papers to the commission for the professor title because, at that point, I was more interested in the research than the title of professor. In Russia, if you are a doctor of medical science and have a job as a professor, you can get a paper that states that you are an official professor.

Chapter 5

The destruction of my Laboratory

On the morning after my dissertation defense, I was at the laboratory as usual.

The neurologist Docent Sakharov was surprised to see me. "If I were you," he said, "I wouldn't work for a year after a dissertation like that."

A vacation was out of the question. I had a tremendous amount of difficult and important work laid out before me. There was also the crucial upcoming meeting with Health Minister B. Petrovsky, for which I had to prepare carefully, putting together my research materials on the anti-edema medication with photographs and sketches.

At the meeting, Petrovsky listened to me and went over everything that I brought.

"Yes!" he exclaimed. "This is real science!"

Right away, he called Zhdanov, the head of the Science Council of Ministry of Health, and asked him to see me. His positive response on my letter: "To Mr. Zhdanov, in the course of three days, please present your arguments regarding the construction of a laboratory especially for Dr. Lerner."

Zhdanov, an organic chemist, quickly understood what I needed and promised assistance. He called a scientist by the name of Gurskoy and assigned him the paperwork for setting up the laboratory. However, the future outcome should serve as a classic example of the wires of bureaucratic entanglement that strangle any new kind of work, even something that was approved from the top. The ensuing indifference and resistance will doom it to failure.

Gurskoy, with whom I met the following day for a serious discussion, was very frank with me.

"Dr. Lerner," he began, "I did not sleep the entire night, thinking about how to help you. I hope you realize that ministers are ministers, but this type of work is done by an apparatus, and that instrument is us. We can speed up the decision, and we can slow it down, and we can set up either a good laboratory or a bad one. Now, everything will depend on you. You can help me too. I am looking for a topic for my doctorate. Your work is exactly what I need. I am being very frank with you here, and I hope we can negotiate something."

I agreed to help Gurskoy since, anyway, I had no other choice. Dostoijevsky once made a distinction between a German and a Russian bureaucrat. If there is an opportunity to permit or restrict something without breaking a law, the German official would choose to permit it, but the Russian would always restrict it.

Gurskoy could not carry out the minister's decision, since he was not a neurologist. He called on another official, the neurologist Inspector Konovalov, for help. Konovalov then called me at my home and told me that he would like to speak with me in confidence and that this discussion should be held as soon as possible. *Poor man*, I thought, *he probably hasn't slept the whole night.* I was anxious to finally place this issue in its right place. In the morning, I went to see Konovalov.

"Dr. Lerner, do you realize that our entire system is a mess!" Konovalov began. "I have a friend at the Central Committee named Shirinsky, who is the director of Public Health. That name is familiar to you, of course. His word is a command! Shirinsky and I will quickly set up the laboratory for you. You will have some coworkers, tools, and currency, and you can do all the work you want. I realize the importance of your research, and I will do all of this under the condition that you will take me as your deputy."

I was a little confused. It is one thing to "share" your research for someone else's dissertation, and it is completely another, when the other person understands absolutely nothing that is presented in the laboratory. To accept this type of condition was impossible. To reject it would mean facing off against Konovalov himself as well as all of his associates.

I urgently had to speak to someone knowledgeable. I called Harlap, who worked at the State Science and Technology Committee overseeing medical research, and knew this system perfectly. He spoke out clearly.

"You do what you want but keep in mind that if you hire Konovalov, he will take over in a wink of an eye," Harlap said. "It is dangerous to get involved with these types of people. The sooner you realize it, the better for you."

He could not have spoken more clearly. I turned Konovalov down.

"Fine, we'll set up your laboratory," he promised, sounding almost angry.

Three days passed, followed by a month, then three months, then six months. It took a tremendous effort to get eight full-time scientists and at least a little bit of money. The institute received an order to set up a laboratory specifically for Lerner. A letter was sent to the Public Health Ministry asserting the importance of my research, but Dean Kovrigina would not sign it. A professor named Shultzev told me why.

"You did serious research, but you did not create this new medical tool alone," Shultzev said.

"Of course not," I said. "There's a chemist who helped me a lot. He works at another institute."

"Which one of our professors, then, took a part in this research?"

"None of them."

"No, Dr. Lerner. Things are not going to proceed in this manner. We have left the days of lonely geniuses behind. What if your work was nominated for the Lenin Award? You cannot get this kind of prize alone, do you understand? You shouldn't have any illusions about this. You need a co-author. Whom do you have for coworkers?"

I sank into my chair.

"I have no coworkers, only a graduate student."

"His last name?"

"Marchenko," I said.

"Wonderful!" Shultzev said. "Marchenko should be your co-author, and his last name is just right!"

"He is not at all affiliated with this work," I objected.

"It doesn't matter. He is not affiliated with it today, but tomorrow he will be. If you are stubborn, Kovrigina will not sign the letter at all."

Marchenko himself was a decent guy. I supervised his dissertation on EEG, which he was to present at the Dentistry Institute. I was warned that because he might have difficulties defending his thesis, it was necessary to include a second consultant from that institute. I was given the name of Professor Smirnov from the neurology department. I felt sorry for Marchenko, for he was actually taking a risk with his own dissertation. I agreed that Smirnov would be the second consultant.

Then, in the written abstract of the dissertation, I saw that Smirnov was featured as the professor and I, Lerner, as the assistant. At the time of the dissertation defense, it was announced that Marchenko's credit would go to Smirnov as the leading professor. This was not the last time I would be forced to add a different "co-author" for the same reason.

The letter signed by Kovrigina stated that the anti-edema medication had been invented by Marchenko and Lerner, with the names listed in that specific order. Only then did I receive a green light for eight positions and a contest for the assistant head laboratory position.

The Communist Party secretary of our department was an insignificant man by the name of Metz. We came to the lab at the same time, but a year later I became the head scientist and he remained a laboratory assistant. On the other hand, he rose in the Party hierarchy. After Metz became secretary, he did not conceal his hostility toward me, caused by his jealousy about my dissertation while he stayed a regular doctor. When he heard I was in the running for the job of head scientist, he decided to take out all of his malice on me. A candidate was to submit a package with a character reference, signed by what was called the "triangle": the director of the department, the secretary of the Communist Party, and the secretary of the labor union. Professor Chetverikov wrote that I was of "very good character," but Metz refused to sign. Without his signature, the reference did not exist, though I myself was not a Communist Party member.

This was a prelude to a completely obscene contest that the bosses started for the position of the director of the lab that was to be dedicated to my research. When I finally got my documents together, I ran into a certain woman named Dobrzhanskaya. About two years earlier, I practically saved her doctorate. At the defense the institute wanted me, as the opponent, to give her a hard time, but out of principal I stayed positive about her dissertation because I liked it. With my help, the Science Council from the Endocrinology Institute approved her dissertation "conditionally," giving her two weeks to correct mistakes. "Eduard, you saved me, I will never forget you!" she told me after the defense. She "thanked" me indeed—by applying for my position.

The destruction of my Laboratory

Now, I needed to hire workers for my laboratory. One of the first I wanted to invite was Dr. Lapitsky. He was my graduate student, was working on his dissertation, and was well acquainted with my research and all of the methodology.

"Lapitsky!" Kovrigina said suddenly.

"Michail," I replied.

Her face flushed in outrage. "I will not allow anyone named Lapitsky because he is a Jewish man!" This convinced me that she wanted to select another doctor.

"That is a strange reaction for a former public health minister," I said. "I'm responsible for this program, including the positions of scientists and coworkers of my department, and I want to work with scientists, not with mediocrities, even if these mediocrities are honest Aryans."

"We still don't know who will be in charge of the laboratory," she said. "I will definitely call Mr. Zhdanov, and if he keeps pandering to you, I will go as far as the Central Committee. I will not allow my institute to become a place for your little cabals."

This was no longer an overture but the main theme being played out aloud, deafening the entire country. In this situation no one was prepared to reckon with the normal rules of simple decency.

The ruling commission usually had meetings to discuss the candidate seven to ten days before the Science Council session. This time, the commission met for about an hour before rendering the final decision. Everything was done in such a way as to prevent me from launching my dissertation. Pale as death, Professor Chetverikov informed me that the commission offered Dobrzhanskaya's candidacy for the head of the laboratory. The outraged professor tried to explain to the colleagues that Dobrzhanskaya was an endocrinologist, not a neurologist, and did not know what a stroke was; she was utterly unqualified to run the laboratory. This fazed no one. Chetverikov then announced, "If Dobrzhanskaya is selected, I will submit my resignation!"

The situation bordered on scandal. The head of the department that created the laboratory recommended a specialist, but a complete non-expert was being squeezed in. In the end, the final commission selected no one, neither me nor Dobrzhanskaya.

Chetverikov submitted his resignation. A person who decides to leave this type of position is placing an "X" on his career, an unbelievable act! Chetverikov's resignation created a huge shock at the institute. This was a blow for Kovrigina and constituted a full revolt at her institute. She tried to persuade Chetverikov that it wasn't worth leaving the department "because of Lerner." She tried to get others to speak with him, but Chetverikov was not to be moved. He appeared to be a man of honor, an even rarer thing in those times than a man of courage.

Soon afterwards, Chetverikov's position was filled by Petelin, the institute's Party secretary. He came to the institute at the same time as Chetverikov. This man thought he was a big expert, but in fact he was struggling with neurology.

When he was elected Party secretary, he came to me in the laboratory and said coldly, "So, Eduard, are you going to do my thesis or not? If you are not, I do not envy your future at all. We will drive you insane. So I suggest you carefully think this over and do my thesis."

I rejected his offer and then, out of the blue, he wrote his doctoral thesis. He copied text from school textbooks and other books and faked medical histories. I even knew the articles from which he had plagiarized entire pages.

How did he become the head of the department? There was a Marxism-Leninism section at the institute, headed by the deputy Party secretary, whose name I cannot remember. In fact, he was in charge of everything at the institute, just not Petelin. Petelin merely got him drunk. As a Party boss, that man could do anything, so he pushed Petelin to head one of the most important departments in the country, to which specialists in neurology from everywhere came for continuing training. Now the head of that influential department was an ignorant impostor who did not even understand his own doctoral dissertation. It was called *Physiology and Function of Hyperkinesis*, but I doubt he could even distinguish hyperkinesis from hypokinesis.[1] Nor, apparently, did he know that "physiology" and "function" meant the same in this context.

Kovrigina went to see Zhdanov, to let him know that as long as she was in charge of the institute, I could not set my foot in the laboratory. Minister B. Petrovsky, who apparently forgot his personal resolution, did not want to endanger his relationship with the former

---

[1] Hyperkinesis refers to excessive violent movements associated with certain neurological diseases. Hypokinesis refers to the limitations in volume and speed of movement due to neurological diseases.

minister and supported her. As the Russian proverb says, "a crow will not poke another crow in the eye."

One person did show interest in my research and decided to help me. His name was Dmitri Fedotov, the director of the Institute of Psychiatry. Instantly, Konovalov and Gursky woke up. I don't know what they whispered in Zhdanov's ear, but his attitude toward me changed abruptly, and he declared that the laboratory (which then existed only on paper) was to be closed down before ever having opened. In vain I tried to convince the Science Council of the Health Ministry, called to review my request, that our division was carrying out an important investigation into strokes, opening huge possibilities for treating many diseases caused by the destruction of the brain's mechanisms. My arguments bounced like a ball off a wall. However, when I presented my medication that had been successfully used in the brain swellings experiment, Zhdanov paused. Words, paper, and dissertation—that's one thing. An effectively built tool, even an experimental one, is a different story.

"That's it," Zhdanov said. "This discussion is over. Tell me, did someone agree to house your laboratory?"

"Yes," I replied. "Professor Fedotov from the Institute of Psychiatry."

"Have him draft a letter today."

Seemingly out of nowhere, Gursky and Konovalov leapt up because both were against me.

"Yuri, what an extraordinary incident!"

"No, an unexpected incident."

"Oh, no, let me tell you."

Zhdanov, however, would not have any of that. "Enough! I know your opinions. The meeting of the committee is over. The decision is that the laboratory will not be closed down and instead will be transferred to the Psychiatry Institute."

I was congratulated, but the fruits of my success were still far away. I had to transfer the entire section of the laboratory to the Psychiatry Institute, including all of the tools, equipment, and so forth. For that, I had to find a new clinic for my laboratory, a very difficult job. On one hand, many hospitals wanted to have a laboratory, which was considered prestigious. On the other hand, because this was new for most hospitals it brought some difficulties. I visited many medical offices until someone I considered a friend recommended Hospital No. 67, where he worked as a proctologist. This was a big, modern hospital on the Khoroshevskoye Road, not too far from the Botkin Clinic and Znamya Truda factory, where I still had many associates. I now had to consult with the chief doctor, Mrs. Petrushko. We reached a tentative agreement, but someone warned me that Petrushko would base her decision on the gift she received.

"Fine, I'll buy her a bottle of French perfume, every woman's dream," I said.

"For this woman the best gift is a color TV set," said my "friend."

Wow! I could not even afford to buy that type of gift for myself. I borrowed a car to move the lab equipment, and I paid for workers, technicians, et cetera. Above all, I had to rustle up some money for Madame Petrushko's gift. It was a very big bribe, a television. I had to hire a couple of "dead souls" (in a scholarly sense, inexperienced students) in the lab, and do it in such a way that they would be able to defend their candidate dissertations. It was a situation worthy of Gogol, though I don't think even he could have dreamt up anything like it.

The gift was accepted gratefully, and the department finally opened, with forty cots for patients. To install the equipment we recieved help from skilled craftsmen from the aviation factory. I had no problem with tools, equipment, or currency. By that time, I had plenty of VIP patients: gas industry Minister Orudzhev, of the Planning Agency; Chairman Baibakov; Lithuania's First Secretary Snechkus; and Central Committee Secretary Peter Demichev, who was in charge of science and public health. I once came to him with a request, but his assistant said, "Peter is very busy. I'll take care of your business. Write only what you need!"

"What is the approximate amount of money you have allotted?" I asked. "It is imported equipment, and we need currency."

"Ask whatever you want. It doesn't matter," the assistant said.

I created an impressive list. The assistant, whose name was Viktor, took the list to Demichev. Soon he was back with a resolution:

"To Minister Petrovsky: Provide equipment immediately."

I think this may have been one of his most expensive autographs. Apparently, he had authorized the purchase of imported equipment worth $200,000!

The destruction of my Laboratory

I wanted to take this priceless piece of paper, but the assistant said that it would be better if they sent the documents themselves to the minister, as I hoped they would.

"Don't worry, Dr. Lerner," he said. "I'll take care of it."

I did not worry. But unexpectedly, Fedotov, the institute's director now in charge of my lab, began to worry. One day he screamed, "Come immediately! What have you done?"

"What is the problem, Dr Fedotov" I asked.

"Come immediately," he yelled, "or else I will have a heart attack!"

It turned out that Minister Petrovsky had called him up in a rage and screamed that he would "destroy us and burn us to a crisp"!

"How did you get to see Demichev? Who the hell are you?" screamed Petrovsky. "Demichev ordered that we urgently buy equipment for Lerner's laboratory. I have barely enough currency for the entire Soviet Union! I cannot fill the secretary's request, and I don't have the right to not fill it. Do something, or else I will destroy you!"

It took me quite an effort to calm Fedotov down. Of course, it was wrong to place him in that situation. Petrovsky still promised to deliver everything that I needed, just not right away. I had to call Demichev's assistant and tell him that everything was in order, the matter was settled, and to give a big thank-you to his boss. Fedotov called up Petrovsky and calmed him down a bit. Apparently, he managed to let off some steam. Even then, however, I didn't know how vindictive Petrovsky was and how loath to forgive. From that moment, the minister became our bitterest enemy.

Meanwhile, I was putting together the laboratory and a section at Hospital #67 with the most modern tools and equipment. Not one laboratory university or clinic in the country, or even the world, had equipment this perfect.

At that time I worked with Hewlett-Packard, a famous American company. HP studied some of my publications and were interested in visiting me in my lab. After their visit, they were so enthusiastic that they promised that we would work together on some problems. I have included HP's original letter informing the Soviet government about the uniqueness of my lab. According to them, there were no more than three or four labs in the world comparable to mine.

Longevity and Health

Just like a stamp collector counts his valuable possessions or an art collector his paintings, so I, too, want to list the equipment that was in the laboratory. All the equipment was manufactured by the best companies: Hewlett-Packard, Siemens, Toshiba, Vickers, Beckman, Nihon Kohden, Nuclear Chicago Corp., Olympus, Hitachi, Kodak. So the doctors among my readers can imagine the remarkable level at which my laboratory operated at that time, forty years ago, I give some examples:

- X-ray machine for angiography;
- Gamma isotope chamber for researching cerebral blood flow and diagnosing brain damage;
- Seventeen- and thirteen-channel EEG machines;
- Eight-channel polygraph;
- Electromiograph;
- A number of microelectronic machines (amplifier, recorder, stimulator);
- Monitor for four patients with automatic recording of EKG, pulse, breathing, blood pressure, and body temperature;
- Telemetric machine for distance recording the above parameters;
- Two-channel biophysiograph with AC-DC amplifier;
- A machine to measure the pressure in the central artery of retina;
- Tachoechoencephalograph;
- Fibrillator;
- Retinal camera for researching and photographing the pupils;
- Stereo static device;
- Pressure chamber;
- Combitest;
- Microastrup;
- Gas analyzer;
- Atomic adsorption spectrophotometer;
- Tromboelastograph; and

The destruction of my Laboratory

- Fluorescent spectrometer.

My biochemical laboratory had even more equipment and was the best in the Soviet Union. The Soviet Union imported many apparatus especially for me because I was a doctor for the Russian government. These highly specialized machines had been developed by great scientists.

In addition were the tools, equipment, and instruments in the neurosurgical unit and reanimation and intensive therapy wards that I set up. There was also a Hewlett-Packard computer that gathered the data from all of this gear, processing the patients' vital data and sending it to a high-powered IBM at the Control Problem Institute, headed by Professor Trapeznikov. This was very special for that period, around 1965–1970. The data processor was supposed to deliver a diagnosis and recommended treatment for each individual situation. This was not just a laboratory with a medical section, but the only one with a completely automatic system for diagnosis and treatment of strokes.

Very soon the laboratory staff grew to forty people: scientists, doctors, engineers, and laboratory assistants. The Stroke Department at Hospital #67 grew in size as well, employing sixty specialists; all told, I was in charge of one hundred people! This did not include the experimental laboratory at the Psychiatry Institute, where research was done on animals. That laboratory was equipped with a pressure chamber (invented by engineer Dalago and myself), electronic devices, and modern equipment.

To no surprise, our automated medical complex elicited admiration from Western visitors, scientists, and medical professionals, and envy from Russian colleagues. We became a thorn in the side of many; they were irritated by our very existence. Compared to that of other laboratories, the quality of our research exceeded the achievements of entire universities, and new methods for treating strokes and some diagnostic methods for illnesses of the nervous system were examples of positive research outcomes that left my laboratory.

Scholars from different countries came to visit me—Austria, Germany, Japan, and others—to work with me and to learn our methods. When the Japanese were developing Anginin, a medication for treating vascular diseases, my laboratory was the only place in the entire Soviet Union to which they sent their instruments for testing.

Longevity and Health

The ranks of our enemies grew, too: E. Schmidt, director of Neurology Institute; N. Bogolepov, a leading neurologist in the country. Both of them had won Heroes of Socialist Labor awards. These men's jealousy was more vicious than a bite of a cobra it is possible to cure any animosity except that which is born out of jealousy.

One of Bogolepov's scientific jobs was writing an article titled "About Great Important Communist Theory in Developing Russian Medical Science" for the biggest medical journal, *Korsakov Journal of Neurology and Psychiatry*. Writing such articles shows that he was not able to do anything important in medical science because he concentrated his article on the communist theory instead of interesting medical topics.

Unwillingly, I become a part of big politics. For example, Dzhermen Gvishiani, deputy chairman for technology at the Council of Ministers, and a man with colossal power, treated me very well, and I was extremely grateful for his assistance. Unfortunately, even he could not save my laboratory and the department; the envy and hate of my enemies were too much even for him. Gvishiani asked Prime Minister Kosygin to get involved. He knew that the latter hated Petrovsky, who had unsuccessfully operated on his wife for cancer. When Petrovsky found out about my good relationship with Gvishiani, the minister decided to put an end not only to my laboratory, but to me as well—to bar me from practicing medicine, period.

Minister Petrovsky, who had the complete support of General Secretary of the Communist Party Brezhnev, wasn't too fearful of Kosygin. In Gagra, where my wife and I often vacationed, I was introduced to a Georgian doctor named Sartia, who asked me for help with his doctorate. He was convinced that there would be no problems with his defense. With his Georgian openness, he told me that every year, Central Committee Secretary Kirichenko visited him on vacation, as did Petrovsky. Sartia gave them gifts: diamonds and antique valuables. Then Petrovsky would pass diamonds to Brezhnev and his daughter.

Kirichenko and Petrovsky created an entire physical therapy institute in the mountains, with Sartia as its director. That is why when Kosygin forced Petrovsky out of his minister job, and the documents reached the Central Committee for approval. Brezhnev still ordered that the minister keep his job. Consequently, Kosygin was unable to help me.

The destruction of my Laboratory

Both the Psychiatry Institute and the Hospital #67 were under the authority of the Health Ministry of the Russian Republic, which is why the punitive operation against me was to be led by Health Minister Trofimov. He set up a commission with six professors from Moscow, Leningrad, and Ryazan, who were to deliver the death verdict on my lab.

Instead, they were amazed and delighted by what they witnessed. Even Minister Trofimov became confused. He knew the ammunition used against me and realized that my enemies were at the top echelons of power—Politburo. He went to seek advice from G. Voronov, a Politburo member. Voronov ordered that my lab be destroyed, and never mind Kosygin, whose enemies were growing in number (Brezhnev, Suslov, Voronov, and others).

This was the crossfire in which I found myself. It was obvious that nothing could save the laboratory—not the intercession of my VIP patients, not even Hewlett-Packard's request to the Soviet government to preserve the laboratory as the only one in the world for diagnosing and treating strokes.

The verdict was a foregone conclusion. Actually, I was given the last word on the matter at the meeting at the Ministry. Like a true philosopher, Gvishiani sounded aphoristic: "Doctor, the actual powerbloc is crying for blood. There's nothing you can do. Just sit back and listen to everything you are being told. If they accuse you, do not try to justify yourself, and if they try to put you down, don't pay any attention. Try to survive the meeting without a stroke, and come and see me immediately afterwards."

I will not go over my entire speech. Basically, I gave a demonstration of the tool we created that allowed research to be done on spinal fluid, as well as the study of meningitis and other diseases of the brain. It could determine the type of swelling, and so on. Minister Trofimov and Shirinsky, the Central Committee member in charge of health care, turned slightly pale.

Everything had been prepared for destroying the laboratory and for destroying me as a doctor and as a scientist. And yet suddenly I was showing that what I had was not the emperor's new clothes, as they maintained, but an exceptional tool like nothing else in the world. The deputy minister for personnel suggested not only that the laboratory be closed, but that I should be banned from doing research and practicing medicine.

Trofimov summed it up: "The laboratory is to be closed, but if Dr. Lerner decides to continue his research, he can do so where he will have proper conditions. We have no objections, and we won't go into the issue of his practicing medicine."

By the time I reached Gvishiani's office, he had already been informed of the verdict.

"If you want to stay healthy, leave the public health system," he told me. "In this mess, you will not be able to do anything anyhow. Settle somewhere where you will be able to do research and work in peace without wasting your life battling the system."

My laboratory was on its deathbed. The following day trucks arrived that could best be described as destroyers. Cutting-edge equipment was loaded like scrap metal, as they tore apart wires, pipes, and everything else. The American-made gamma chamber was sent to Leningrad.

Having found out about all of this, radiology Professor Zubovski fought for a way to convince Minister Petrovsky to leave the equipment in Moscow. The train stopped at a station in order to reload the chamber into another car. Subsequently, all the optical and electronic equipment was stolen en route, so that by the time the machine reached the Roentgen Institute in Moscow, all that was left was the metal body. American specialists were called in to rebuild the equipment. They were horrified to see what was left. The lab was gone.

My friends and associates were surprised at how I could handle such stressful situations without having a heart attack or a stroke. What saved me was my faith in the importance and success of my work, as well as daily physical exercise, which is known to reduce stress.

Back in Moscow, Gvishiani advised that I find a quiet place where I could work peacefully without wasting my energy on fighting. He considered a war on the system useless. His second bit of advice was that, in order to keep my health, I needed to stay as far away from the Health Ministry as I could. The most important warning should have been written not on packs of cigarettes, but in huge letters on the façade of the Ministry: *The Minister warns you that he is harmful to the health of the citizens.*

"Do you have anything else in mind? Where would you like to work?" asked Gvishiani.

"Djermen," I said. "I heard that Professor Parin set up a laboratory researching the functions of humans and animals. The

The destruction of my Laboratory

laboratory works under the aegis of the Academy of Sciences. I would like to work there."

Gvishiani immediately called Millionschikov, the vice president of the academy, and asked him to give me a job, hinting that the request came not just from him, but from "you know who," et cetera. Millionschikov did not need any explanations; he could read between the lines and knew about the Kosygin controversy.

"Eduard, did you hear everything I said? Go see Millionschikov immediately. He will see you."

It was not easy to see the academy vice president, but Millionschikov received me. He asked me where I wanted to work.

When I told him, he said, "Well, if Professor Parin agrees to take you, I will sign the request for you to get a position at the laboratory as a senior researcher."

My request stunned Parin. An even bigger surprise was my assurance that his laboratory had a vacancy as a senior researcher that he didn't know of. He knew that getting that position was almost impossible. He could not reject me in front of the president, and Gvishiani's request was quite valuable. Parin agreed to take me and asked to be introduced to Gvishiani as well. Perhaps he wanted to be sure that I wasn't some kind of con man. The following day, Gvishiani met with Parin, and the latter was impressed. Gvishiani confirmed his request.

"Comrade Parin," he said, "I am asking that you please hire Dr. Lerner. You can address me with any questions. If you need tools, solutions, equipment, just ask, and I will provide you with everything."

Parin said that the laboratory was new, and it was in need of imported equipment.

Gvishiani understood immediately.

"Let's do this," he said. "Make a list of companies that produce the imported equipment your laboratory needs. We will organize an exhibition right in your laboratory, and I will buy you anything you want. In addition, I promise you a couple of scientists to work for you."

Parin was so shaken that it took him a while to find the door and leave. At the time, he was about seventy, and he seemed a little sick. His face was pale, worn out, with droopy eyes. Few people would be healthy if they had lived through what he had: war, prison, camps (he had been sentenced on the famous "anti-Soviet" Article 58-10, "anti-Soviet and counter-revolutionary propaganda and agitation," which was often used against people who voiced opinions disagreeing with Stalin. It carried a minimum of six months imprisonment.). After release, he had been involved in space research; he headed the Space Medicine Institute, but eventually had to leave and went to set up his own laboratory. This outstanding scientist had many enemies, too. This is why Gvishiani's agreement to assist with the laboratory was so shocking to him.

The laboratory was located in an old, quiet part of Moscow, Pokrovskoe Streshnevo, not far from the Kurchatov Institute. Unfortunately, just like in the world outside, there was no peace inside the laboratory, either. It was split into two camps, Grastchenkovists and Parinists. Grastchenkov had been in charge of the laboratory at one point and had his own followers. I found myself in Parin's camp— he asked me to work with his group—but generally distanced myself from discord and quarrels.

One of the leaders in the laboratory was Professor Gecht. He saw me as a rival, though I gave him no reason to and set a good number of coworkers against me. He kept getting on my nerves, but I do not wish to reminisce on all of his plotting.

After Parin's death, in 1971, the laboratory immediately split up. Those who had been employed the longest were transferred to the Institute of Physiology, outside the academy, while the more recent hires, including myself, went to the Severtsev Evolutionary Morphology and Animal Ecology Institute, led by the academician Vladimir Sokolov. I spent almost twenty years there, until I left for Holland, doing research on the autonomous nervous system and paying special attention to skin potentials meaning research on the electric fields in the skin.

My group wasn't big at all, and I was helped by a doctor by the name of Bondarchuk. He had once worked for me at the Psychiatry Institute but had disappeared after the destruction of the laboratory. There were rumors that he defended his dissertation in the Institute of Surgery with Petrovsky. We met again afterwards, and I hired him. We studied cardiography, the curve components that are regulated in the autonomous nervous system.

I handled the medical part of the research, and Engineer Chernyshov worked with data on the computer, using his own method. However, not everything that is new is effective. To verify his method, I took a round piece of dishware, traced its circumference with a pen, and presented this "curve" to Chernyshov. He had produced the data, and in his words, it was "just fantastic." After discovering that, instead of an EKG, he had calculated a soup bowl, the professor got mad at me and for a long time wouldn't forgive me. He had to ditch his method but tried to get even with me every chance he had.

I no longer did research on treating strokes, but instead studied mechanisms that caused apoplexy due to hemorrhage in the brain. One of the main causes of strokes is blood pressure. The autonomous nervous system plays a major role in the regulation of the blood pressure. It was very hard to get into the tiny mechanism of the autonomous nervous system, and I tried all possible methods, including EEG, cardiography, rheography, and plethysmography, without success. A completely new approach was needed for the study of the nervous system.

The search for the new method went on for many years. What I was searching for seemed as elusive as a neutrino. Then one day, by chance, while recording skin potentials, I discovered that some people's skin potential had very high amplitude, while others were missing all fluctuations entirely. I started to collect the data taken from hundreds of people, both healthy and ill.

An old theory in medical science holds that the autonomous system is divided into two parts: the parasympathetic and the sympathetic. This old theory from the 1920s had been consigned to oblivion, and studying it was very time-consuming. Autonomous neurology followed the path of studying specific autonomous regulations. From my tests, however, it appeared that people with high-amplitude fluctuations were parasympathetically active, and those who lacked fluctuations were sympathetically active. My new method to easily diagnose this distinction between the two parts of the autonomous system was a big breakthrough. In Chapter 12, I will elaborate on this new invention, called EAG.

My return to the old forgotten theory about the autonomous nervous system was regarded as odd. People did not see that I was taking it to a higher level and using old notions in a completely new context. After a couple of years of research, I was able to find a method later called EAG, publish it in the Soviet Union, and receive my author's certificate.

Longevity and Health

One day in the summer of 1971, I arrived at my laboratory to discover that my office had been ransacked. The tables were overturned, and all the papers from the drawers were on the floor. I was told that a couple of men had broken in to wreak havoc. I soon found out who did this.

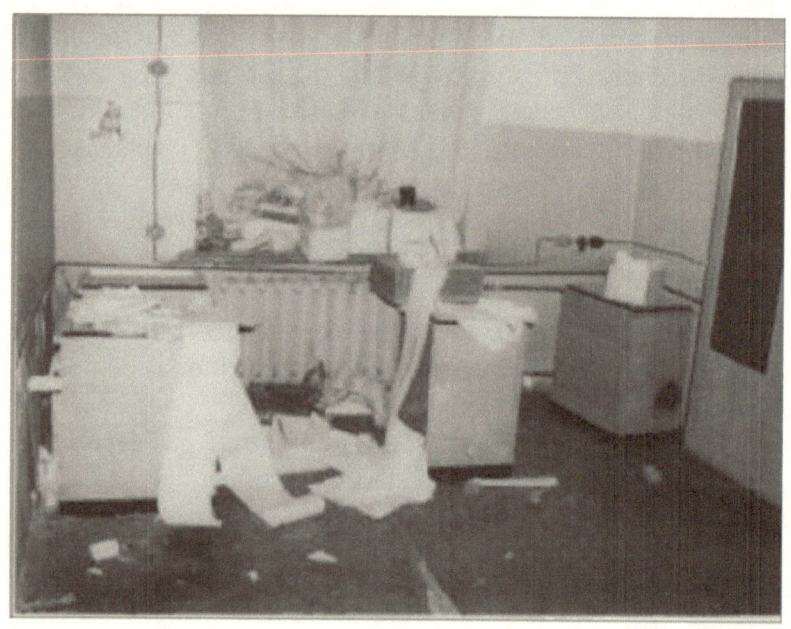

The destruction of my office in Moscow.

After some time, I discovered that some material on the autonomous nervous system was missing. How fortunate that the main materials were stored in a different place! Bondarchuk had access to the material because we worked together. When I questioned him about who could have taken those materials, it turned out to be no one else but Bondarchuk himself.

And where did he take them? Home.

Why? I couldn't believe this. He said he needed them.

"What do you mean you needed them?" I demanded. "If you need to work with them in the laboratory, I need to use them also! Return them immediately!" I said.

"I cannot give them to you."

"What do you mean you cannot return my documents?"

42

The destruction of my Laboratory

"I will be needing them for a long time," he answered.

"Have you lost your mind? How could you even dare to take them without permission?"

"I just took them. I cannot tell you anything else," he said.

I went to the director with a request to get rid of this impudent person. What a surprise it was when the director threw his hands up and said, "I can't. Bondarchuk has powerful protectors."

The director of the institute was even more frank: "Eduard, don't get involved. He works with the State Security."

Six months later, I received my documents back from Bondarchuk. Apparently, someone was interested in my work, but who was that someone?

And yet I enjoyed my job. I had finally found a quiet place, as Gvishiani had advised. The job was peaceful and interesting, and I had the freedom of coming to the laboratory once a week or once a month, without anyone demanding my presence. My job was to generate ideas, and this depended totally on my own motivation. It seemed that a new era had dawned in my life. I was able to progress substantially in my research and understand some of the most difficult, tiniest processes and mechanisms of the autonomous nervous system.

In 1989 I told the institute director that I wished to go to Holland for two years in order to continue my research there. Sokolov promised to help and did everything he could. I got the necessary recommendations of the Science Council. I went through the hoops of the Party Bureau and gathered all the required papers. Now they were to be reviewed by the academy and the biology department. Finally, once again thanks to Sokolov, my papers ended up with the Secretary Georgiy Skriabin, the son of Constantin Skriabin, a famous parasite expert—which describes Skriabin very well. He could not believe that I, a Jew, was being sent to Holland for two years, and with my family, too. These types of things did not happen in the academy especially for Jewish people it was very hard.

My documents were then forwarded to the international section of the academy. They put me through the third-degree about my parents, grandparents, and daughter. Finally, they agreed to issue me an official blue passport, different from the common red one. One of the employees of the international academy who by then knew me well—and was my patient, too—gave me valuable advice. He advised me not to accept a blue passport under any circumstances—that if I did, I would be like a dog on a leash. If I ever chose to go from Holland to another country, I would need to ask permission from the government, which would then have to file a request with Foreign Ministry. If, however, I went to Holland on an invitation from my daughter that married a Dutch man, I would be able to move freely wherever and whenever I wished, without anyone's permission.

Although the blue passport had already been issued, I still managed to get a regular one and leave for Holland as a public citizen. It sounds quick, but in reality it took forever, with grave upsets and stresses and constant fear that the door would close in my face and I would never be able to get out of this country. Every slide, every page that I wanted to take with me needed to go through inspection at a certain office that issued the exit visa. Then the permission was verified, re-verified, and confirmed.

Finally, with all the stupid procedures in the past, I was allowed to bring eight transparencies, twelve slides, and eighteen typed pages. In May of 1988, I left for Holland.

On the next pages I present documents to show my professional position at the time I left Russia, followed by pictures from my youth.

As I mentioned earlier, in Russia there are two dissertations: the first a candidate of medicine, comparable to a PhD, and the second a doctor of medicine, which entitles one to the position of professor. I finished my candidate dissertation during the fourth year of the Medical Institute while still a student. Once again, I was one of the youngest ever to complete the doctoral dissertation.

The destruction of my Laboratory

Translated from Russian
copy

HONOURS   DIPLOMA
No.297I08

This certifies that Lerner Edward Naumovich
was admitted to the Chernovtsy State Medical Institute in I946
and that he completed the full program of the said Institute
in I95I majoring as
"general practitioner".
Lerner E.N. has been awarded the qualification of
"general practitioner"
pursuant to the decision of the State examining board dated
July 3,I95I.

Chairman of the State
examining board   - sg
Director - sg
Secretary - sg

The town of Chernovtsy, I95I
Registration No.I085

Seal of the authority which
has issued the document

The City of Moscow, this 4th day of July in the year I989.
     I, Zimina Rimma Ivanovna, Deputy Chief State Notary to
State Notary Office No.I in and for Moscow, hereby certify that
this is a true copy of the original, there being neither erasures
nor crossed out words nor postscripts nor any other unspecified irregularities in the latter.
Registration No.25з- //45

          Duty collected 20 kopecks
Deputy Chief State Notary - signed: R.Zimina
Seal: the RSFSR Ministry of Justice
     Moscow State Notary Office No.I

Translated by Nagornaya O.N.

M o c k -

ва, двадцать шестого июля тысяча девятьсот восемьдесят девятого года.

Я, Соколова Ольга Владимировна, государственный нотариус Первой московской государственной нотариальной конторы, свидетельствую подлинность подписи, сделанной известным мне переводчиком Нагорной Ольгой Николаевной.

Зарегистрировано в реестре за № 243-451

Взыскано государственной пошлины 20 копеек.

Государственный нотариус 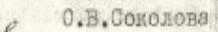 О.В.Соколова

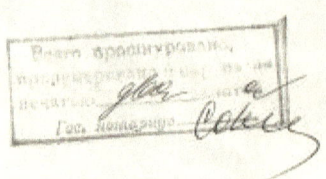

Moscow, July 26, 1989

I, Olga Vladimirovna Sokolova, the government notary, from the first Moscow government notary office, confirm that this is a true document and my signature is official, and the translation is made by O. N Nargonaya, whom I know.

It was registered under number 243-451.

We took 20 kopeek as a government rent.

The Government Notary ….. O. V. Sokolova

The destruction of my Laboratory

Translated from Russian
copy

U S S R
Ministry of Higher and Secondary Specialised
Education
Supreme Attestation Commission

D E G R E E   O F   CANDIDATE OF
MEDICAL SCIENCES
No.004726
Moscow   May I4,I958

This certifies that LERNER Edward Naumovich
has been admitted to the Academic Degree of
CNADIDATE OF MEDICAL SCIENCES
pursuant to the decision of the Academic Board of the Central
Institute of Further Training of Physicians   dated November I2,I957 (minutes of proceedings No.I6) .

Chairman of the Board - sg
Scientific secretary of the Academic Board - sg

Sealof the authority which
has issued the document

The City of Moscow, this 4th day of July in the year I989.
    I, Zimina Rimma Ivanovna, Deputy Chief State Notary to
State Notary Office No.I in and for Moscow, hereby certify
that this is a true copy of the original, there being neither
erasures nor crossed out words nor postscripts nor any other
unspecified irregularities in the latter.
Registration No.25₃- 1157

                    Duty collected 20 kopecks

Deputy Chief State Notary - signed: R.Zimina

Seal: the RSFSR Ministry of Justice
      Moscow State Notary Office No.I

Translated by Nagornaya O.N.

                                    M o c k -

ва, двадцать шестого июля тысяча девятьсот восемьдесят девятого года.

    Я, Соколова Ольга Владимировна, государственный нотариус Первой московской государственной нотариальной конторы, свидетельствую подлинность подписи, сделанной известным мне переводчиком Нагорной Ольгой Николаевной.

    Зарегистрировано в реестре за № 24з-4505

    Взыскано государственной пошлины 20 копеек.

Государственный нотариус 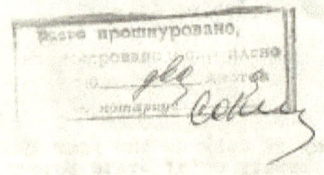 О.В.Соколова

Moscow, July 26, 1989

I, Olga Vladimirovna Sokolova, the government notary, from the first Moscow government notary office, confirm that this is a true document and my signature is official, and the translation is made by

O. N Nargonaya, whom I know.

It was registered under number 243-4505.

We took 20 kopeek as a government rent.

The Government Notary   O. V. Sokolova

The destruction of my Laboratory

Translated from Russian
copy

U S S R
Ministry of Higher and Secondary Specialised
Education

Supreme Attestation Commission

CERTIFICATE OF SENIOR SCIENTIFIC
SPECIALIST
MCH   No.016027
Moscow July 31,1964

This certifies that Lerner Edward Naumovich
has been awarded the academic title of
SENIOR SCIENTIFIC SPECIALIST
pursuant to the decision of the Supreme Attestation Commission
dated July II,1964 (minutes of proceedings No.32/n), his
speciality being
        "neurology".

Chairman of the Supreme Attestation Commission - sg
Scientific Secretary of the Supreme
    Attestation Commission - sg

Seal of the authority which
has issued the document

The City of Moscow, this 4th day of July in the year 1989.
    I, Zimina Rimma Ivanovna, Deputy Chief State Notary to
State Notary Office No.I in and for Moscow, hereby certify
that this is a true copy of the original, there being neither
erasures nor crossed out words nor postscripts nor any other
unspecified irregularities in the latter.
Registration No.25₃- //63

Deputy Chief   Duty collected 20 kopecks
State Notary - signed: R.Zimina
Seal: the RSFSR Ministry of Justice
    Moscow State Notary Office No.I

Translated by Nagornaya O.N.

                                        M o c k -

49

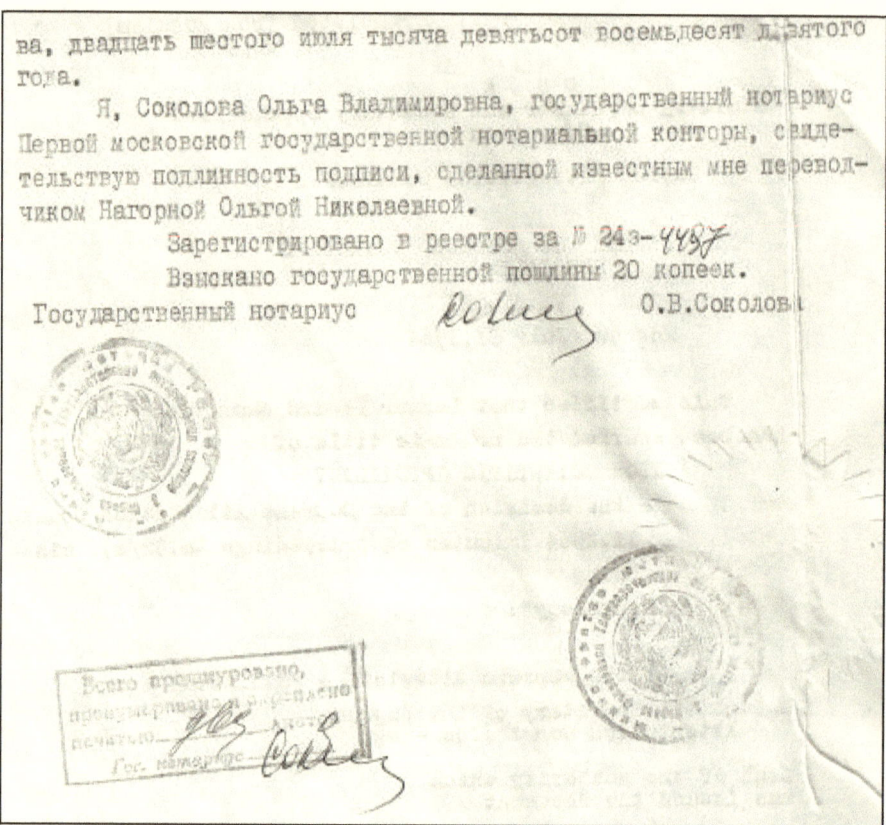

ва, двадцать шестого июля тысяча девятьсот восемьдесят девятого года.

Я, Соколова Ольга Владимировна, государственный нотариус Первой московской государственной нотариальной конторы, свидетельствую подлинность подписи, сделанной известным мне переводчиком Нагорной Ольгой Николаевной.

Зарегистрировано в реестре за № 24з-4497

Взыскано государственной пошлины 20 копеек.

Государственный нотариус   *Roluey*   О.В.Соколова

Moscow, July 26, 1989

I, Olga Vladimirovna Sokolova, the government notary, from the first Moscow government notary office, confirm that this is a true document and my signature is official, and the translation is made by

O. N Nargonaya, whom I know.

It was registered under number 243-4497.

We took 20 kopeek as a government rent.

The Government Notary   O. V. Sokolova

The destruction of my Laboratory

U S S R
Ministry of Higher and Secondary
Specialised Education

D E G R E E   OF DOCTOR OF MEDICAL SCIENCES
No.001243
Moscow March 8,1965

This is to certify that Lerner Edward Naumovich
has been admitted to the academic degree of
DOCTOR OF MEDICAL SCIENCES
pursuant to the decision of the Supreme Attestation Commission
dated February 6,1965 (minutes of proceedings No.5).

Chairman of the Supreme Attestation Commission - sg

Scientific Secretary of the Supreme
Attestation Commission - sg

Seal of the authority which
has issued the document

The City of Moscow, this 4th day of July in the year 1989.
I, Zimina Rimma Ivanovna, Deputy Chief State Notary
to State Notary Office No.I in and for Moscow, hereby certify
that this is a true copy of the original, there being neither
erasures nor crossed out words nor postscripts nor any other
unspecified irregularities in the latter.
Registration No.25₃- 1161
Duty collected 20 kopecks
Deputy Chief State Notary - signed: R.Zimina
Seal: the RSFSR Ministry of Justice
Moscow State Notary Office No.I

Translated by Nagornaya O.N.

M o c k -

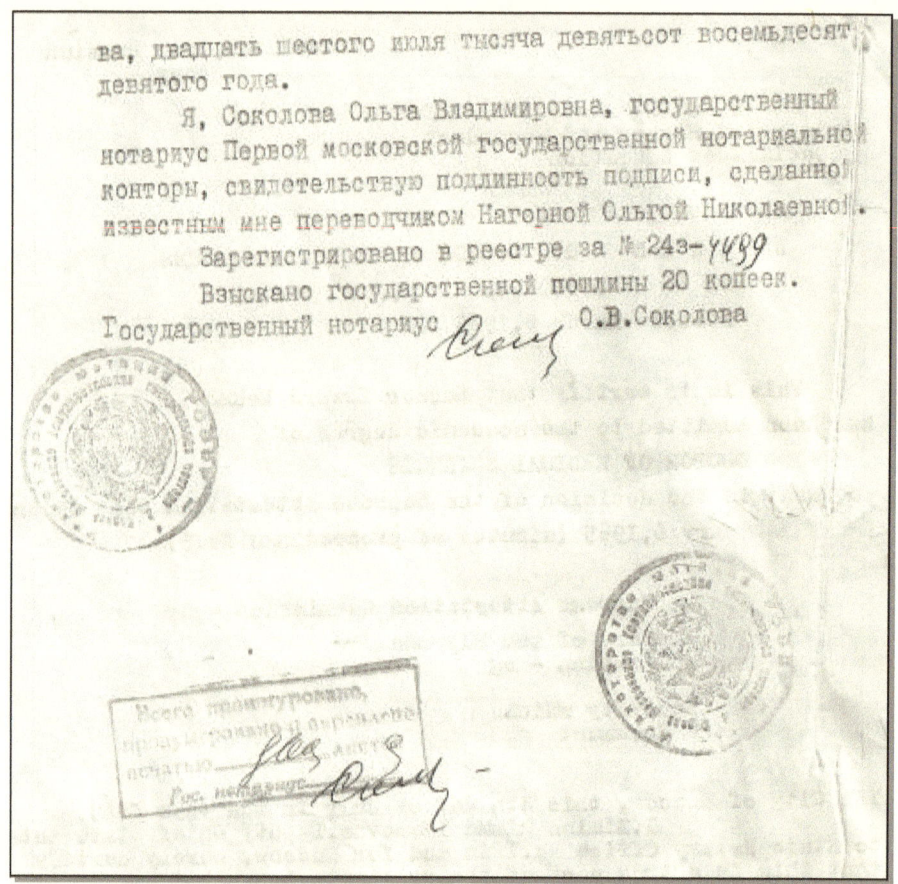

Moscow, July 26, 1989
I, Olga Vladimirovna Sokolova, the government notary, from the first Moscow government notary office, confirm that this is a true document and my signature is official, and the translation is made by O. N Nargonaya, whom I know.
It was registered under number 243-4499.
We took 20 kopeek as a government rent.
The Government Notary   O. V. Sokolova

The destruction of my Laboratory

REFERENCE
for the Method of Estimating the Condition of the
Vegetative Nervous System Suggested by B.N.Lerner,
Doctor of Medical Sciences

The method of electrovegetography (EVG) suggested by
B.N.Lerner is characterized by its simplicity and availability
for a wide range of physicians in their everyday work. The
method makes it possible to determine the condition of the
vegetative part of the nervous system in many diseases, to improve the treatment and, which is especially important, to make
a prognosis of the course of hypertonic disease with its complications (insults, myocardial infarctions etc.) ulcerous
disease of the stomach and duodenum, dyskinesias of the gastro-
intestinal tract, vegetative dystonias and oth.
    EVG will also be adopted for occupational selection, for
the control of individual medicamentous therapy.
    Convincing examples are given in the claim for invention.
The formula of the invention is well grounded and arouses no
objections.
    I recommend to record this claim as an invention.

State prize laureate, Member of
the Academy of Medical Sciences
of the USSR   Professor L.O.Badalian

Professor L. O. Badalian

3-04-1983

This was translated
from Russian to
English language by

Golemba Nadiasda
Pavlovna
Moscow

г.Моска,двенадцатого марта тысяча девятьсот восемьдесят восьмого года.

Я,Зимина Римма Ивановна,заместитель старшего государственного нотариуса Первой московской государственной нотариальной конторы, свидетельствую верность этой копии с подлинником документа,впоследнем подчисток, приписок, зачеркнутых слов и иных неоговоренных исправлений или каких-либо особенностей не оказалось.

Подпись переводившего с русского языка на английский язык сделана лично мне известным переводчиком Големба Надеждой Павловной, подлинность подписи которого свидетельствую.

Зарегистрировано в реестре за № 253 - *296*
Взыскано государственной пошлины 40 копеек.

Заместитель старшего
государственного нотариуса   Р.И. ЗИМИНА

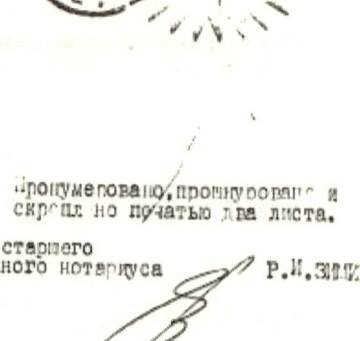

Пронумеровано, прошнуровано и скреплено печатью два листа.

…ель старшего
…твенного нотариуса   Р.И. ЗИМ.

Moscow, March 12, 1988
I, Simina Rima Ivanovna, the assistant of the head government notary, from the first Moscow government notary office, confirm that this is a true document without any changes. The signature from the translator that translated from Russian to English language is official. This is Golemba Nadiasda Pavlovna, whom I know.
It was registered under number 253-296. We took 40 kopeek as a government rent.

The Assistant Government Notary …..
Simina Rima Ivanovna

The destruction of my Laboratory

REFERENCE

for the Method of Estimating the Condition
of the Vegetative Nervous System Suggested
by E.N.Lerner, Doctor of Medical Sciences

The vegetative part of the nervous system is known to play most important part in the pathogenesis of the majority of human diseases. However, in spite of great achievements in different fields of medicine, investigations in the vegetative part of the nervous system are not much advanced. That's why we should welcome method of electrovegetography (EVG) suggested by E.N.Lerner.

The method is simple, does not require any special apparatus and can by used under hospital, dispensary and even home conditions.

EVG can be especially important in prophylactic medical examination of people as a prognostical method of diagnosis at the pre-clinical stage of such severe diseases as hypertonic disease with its complications (insults, myocardial infarctions etc.) diseases of the gastro-intestinal tract and oth.

At present it is still difficult to estimate all the significance of EVG for medicine as it has just begun to develop but in our opinion the method is promising. The formula of the invention set forth in the claim does not evoke any objections.

The claim for the invention is substantiated.

Head of the Chair of Nervous
Diseases of the Therapeutic
Faculty   Professor V.A.Carlov

2-04-1983

This was translated from Russian to English language by

Golemba Nadiasda Pavlovna Moscow

Перевела с русского языка на английский язык
Големба Надежда Павловна
Н. Голем

г.Моск-

55

ьа,двенадцатого марта тысяча девятьсот восемьдесят восьмого года.

Я,Зиминя Римма Ивановна,заместитель старшего государственного нотариуса Первой московской государственной нотариальной конторы, свидетельствую верность этой копии с подлинником документа, в последнем подчисток,приписок, зачеркнутых слов и иных неоговоренных исправлений или каких-либо особенностей не оказалось.

Подпись переводчившего с русского языка на английский язык сделана лично мне известным переводчиком Големба Надеждой Павловной, подлинность подписи которого свидетельствую.

Зарегистрировано в реестре за № 25з - 299
Взыскано государственной пошлины 40 копеек.

Заместитель старшего
государственного нотариуса   Р.И. ЗИМИНА

Пронумеровано,прошнуровано и
скреплено печатью пвх листа.
Заместитель старшего
государственного нотариуса   Р.И. ЗИМИНА

Moscow, March 12, 1988
I, Simina Rima Ivanovna, the assistant of the head government notary, from the first Moscow government notary office, confirm that this is a true document without any changes. The signature from the translator that translated from Russian to English language is official. This is Golemba Nadiasda Pavlovna, whom I know.
It was registered under number 253-299.
We took 40 kopeek as a government rent.

The Assistant Government Notary ..... Simina Rima Ivanovna

The destruction of my Laboratory

Myself at age eighteen

Longevity and Health

*Photo taken in 1949 showing myself with my fellow students
at the Medical Institute in Chernovtsy in the Ukraine.*

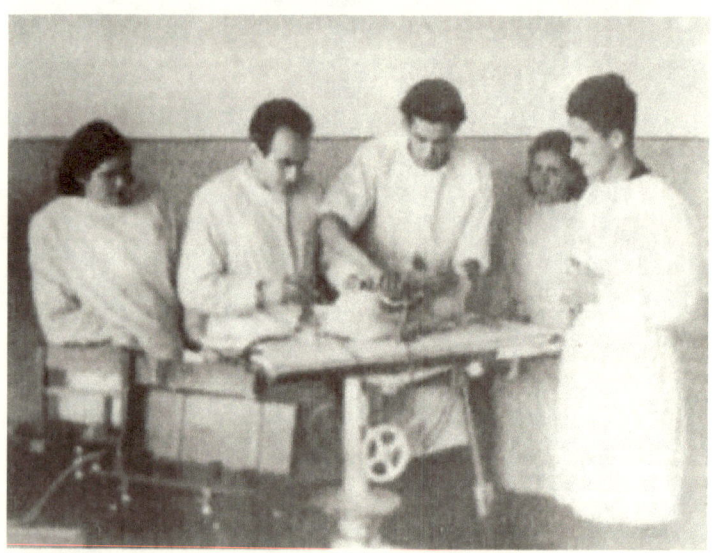

*Photo in 1948 showing myself experimenting on a dog in the
Patophysiology Department with the assistance of two doctors, a laboratory
assistant, and a friend (Resnik, who later became a professor of surgery).*

The destruction of my Laboratory

Myself and engineer Dalago with our invention, the Barocamera. This device can make the oxygen pressure level in the human body much higher. That makes it easier to treat patients with strokes or other brain diseases.

A dinner I gave after my dissertation in 1963. Professor Narikashvili is seated in front of me. My wife is seated to the right of Professor Narikashvili, and next to my wife is Professor T. Jachenko, the head of a big laboratory in Moscow. In the foreground are my colleagues.

Chapter 6

Scholarly Theft

After chapters so substantial, I will now turn to miscellaneous, short notes I made over the years. They deal with certain qualities that were, unfortunately, typical of a number of my colleagues: ungratefulness, lack of integrity, envy, and enmity.

I first ran into what I term scholarly theft in my third year at medical school. I have already written about my days spent in the Pathological Anatomy department, dyeing bone marrow nerves. I did a general description of the method of the whole process and obtained bone marrow preparations that contained unbelievably beautiful nerves. I could not show them in Chernovtsy, however, since there were no nerve fiber experts there.

There was a professor's assistant, Borima, who worked in obstetrics. One day, he was going to Kiev on personal business; I asked him to take my preparations along to show them to an expert in Kiev. When he came back, he told me he had lost the preparations.

After a couple of years, I discovered that a doctoral dissertation on bone marrow enervation had been successfully defended at the histology department in Kiev. That was my first experience of scholarly theft.

During another period, while at Botkin Hospital, I studied the effect of solar activity on cardiac illness. Space biology was pioneered by Alexander Chizhevski, who, back in 1917, after processing tons of data, established that in most people high solar protuberances caused higher nerve excitability, emotionality, and excessive movements. An international society for biological meteorology was created in Holland in 1956 to study the relation between the changes in the weather and human health.

As you see, this problem wasn't new but, we had our own approach and our own original ideas. Together with some of my employees that I selected for this job, I opened Moscow's emergency service archives, going back many years, and tracked the number of strokes and heart attacks on each day. A lab at the Energy Institute studied solar activity. We correlated our research, and the results were impressive. In order to produce valid research, we needed to enter the entire volume of data into a computer.

I turned to Professor Studnitsky at the academy's computing center, who sent me to see Professor Zhuravlev, who, he thought, could make these calculations. Even a cursory look at my data told Zhuravlev there was a high probability that cardiac disease was related to solar activity. He promised to tell me the results in about two weeks.

Weeks passed, even months, but I simply could not reach him. He did not call me, and I could not get him on the phone. He was completely unreachable. I have never seen him since. Professor Studnitsky was of no help either. However, shortly afterwards, newspapers ran articles warning people with vascular heart problems about critical days with increased solar activity and advised them to follow certain doctors' advice. From this time on, such warnings became common. Apparently, Professor Zhuravlev used the material he had stolen from me in order to become the corresponding Academy member.

In the beginning of the 1960s, I had a chance to study electroencephalography with Professor Zhirmunskaya at the Institute of Neurology. I once shared with her my idea to create an EEG atlas for strokes, combine EEG with strokes, and thus develop stroke diagnostics. The professor reacted to my idea coldly, yet compiled a similar atlas herself, a year and a half later. Just like that!

Another example occurred with a young professor named Portnoy, who worked at Moscow's Vishnevsky Surgery Institute. At Botkin, I studied diagnosis and treatment of strokes, at which time I discovered a very interesting fact. The case was so striking that I told Portnoy about it. When the professor did not believe it, I described the details. After some time, *Surgery Magazine* ran an article by Professor Portnoy describing that very same fact. Of course, my name wasn't mentioned or referenced at all.

The problem of plagiarism in scholarly work is not a new one Yet every time one stumbles upon it, he feels like a robbery victim. I can

Scholarly Theft

recall many ideas that my colleagues have stolen from me. I have to note that scholarly theft is widespread not only in Russia but in the West as well. Regardless of the language spoken, I have had to learn to know whom I can trust and who is capable of deceit and dishonesty. One has to remain silent, which makes it very hard or even impossible to work with colleagues, while realizing that the results of the research can be taken by a dishonest worker.

The following story is not so much about plagiarism but simple ingratitude. In the late 1960s, as I wrote, I worked at the Botkin Hospital; more specifically, I was in charge of the stroke section at the Central Institute for Continuing Education. Professor Chetverikov, who was in charge of the department, asked me to help a Dr. Berginer from Kishinev with his doctorate. The professor introduced me to Berginer, who told me that he had completed his internship at the Neurology Institute and had basically wrapped up his dissertation there. He needed only to fix it up a little in order to be ready for its defense.

I was interested in reading his work, because the Neurology Institute was a leader in its field. I took the dissertation home and was shocked to see that, though the subject concerned treating the disruption of cerebral circulation with IV Euphilin, only one case was cited in the entire work. The following day, I asked Berginer how he could write a dissertation based on one single incident. He murmured something under his breath.

It wasn't just anybody who had asked me to help; it was Professor Chetverikov. On the other hand, I had plenty of stroke patients in my care—after all, this was our specialty. I decided to help this Kishinev doctor. I turned the entire department's resources over to helping him. I did his EEGs, I tested different methods, and in a month, I amassed a decent amount of material. I also had to edit his entire work, because it wasn't written at the required level.

When everything was completed, he went home. To this day, I do not know where he defended his dissertation. Usually, if a dissertation was done in my department, either Professor Chetverikov or I became the adviser. In Berginer's case, no one knew who the adviser was, where the dissertation was defended, or whether he succeeded.

I later found out that he had picked Chetverikov for his adviser. Not only did he not invite me to the reception after his dissertation, but he even neglected to thank me for my assistance. In fact, he got his entire dissertation done in our very own department.

Later, Berginer left for Israel. We lived in Holland at the time, and one day my wife and I went to Israel for a few days and found ourselves in Beersheba, where a friend of ours worked at the local medical institute. She mentioned Professor Berginer during our conversation. Verifying his name and profession, I realized that it was the same Berginer who had done his dissertation in my department.

Earlier, I read in the paper—or someone wrote me in a letter— that Berginer lived in Israel and had told everyone that he was the student of Professor Chlenov and Academician Schmidt and that he had completed his internship and dissertation at the Institute of Neurology. Chlenov and Schmidt were leading neurologists, and it was an honor to study with them. However, his report of doing his dissertation in the Institute of Neurology just did not jibe with facts.

I called Dr. Berginer, and without asking any questions he invited me over. My wife, my friends, and I went to his house. He put on tea, and we sat around and chatted.

Suddenly, he said, "Dr. Lerner, I cannot remember where I met you before. Was it in the Institute of Neurology, or in the library while I was collecting my research?"

I was taken aback. At first, I thought that maybe he was suffering from Alzheimer's; I began to remind him slowly of the associates, professors, and doctors we both knew. He remembered everyone perfectly: Chlenov, Schmidt, Tsuker, along with a few others. I became sure that it wasn't Alzheimer's; it was a completely different illness called lack of conscience. I chose not to point it out to him. After tea, I got up and so did my wife and my friends. We said our good-byes and left. This was my last encounter with the person whose dissertation I had practically written.

Although there are many other examples of this kind of behavior, I will give only one more. In the Moscow Botkin hospital where I worked for many years, there also worked a Dr. Shtock. He was a translation specialist, translating from English to Russian, and he published short versions of the translations in Russian medical journals. At that time, everybody was sure that he knew excellent English. Twenty years later, I called him from Amsterdam and invited him to a symposium. To my surprise, he told me he didn't understand

Scholarly Theft

English! How was it possible that he had published these English translations in Russian journals?

Recently, I called him and asked him if he could collaborate on a project with me. He wasn't very valuable to me, but he was one of the last scientists that I knew in Moscow who was still working in this area. To my surprise, he told me that he is now a professor and the head of the neurological department at Botkin Hospital. He was not nice to me and rejected the idea of doing anything with me. He had forgotten that I had opened his path to science by helping him with his dissertation.

To be fair, I must tell you about another colleague, named Davidovich, who lives in Israel, studied with me in Chernovtsy. It turned out that he went to work in Lutsk as a neurologist, and I ended up in Moscow. He went to graduate school by correspondence, and Chetverikov suggested that I should be his advisor. I agreed. Davidovich did a doctoral dissertation under my supervision, successfully defended it, and ended up in Beylenson Hospital in Israel.

Now, when I visit Israel, he welcomes me. Not only does he remember everything but also sincerely thanks me again for helping him complete his dissertation. What this all says is that some people do not have a conscience, and some people do. Despite my bitter experience, I continue to hope that the latter are in the majority. However, in the following chapters, I will recall once again some outrageous examples of lack of scholarly integrity—not only by doctors, but by big companies as well.

Chapter 7

My Stamp Collection

In the early seventies, I lost my laboratory and department in the hospital. Suddenly and unexpectedly, I had a lot of free time. It was a pity to waste it on crying and complaining about cruel fate. Instead of killing time drinking vodka or playing cards, I decided to use its abundance to expand my interests and thus began to collect stamps.

I got this inspiration from Yuri Slutsker, one of the most famous Soviet philatelists. Since childhood, he suffered from an incurable disease that prevented him from going to school or working. Someone advised the disabled boy to collect stamps. The childhood hobby became his passion, his consolation, his joy, and the meaning of his life. Slutsker is long gone, but his stamps helped him lead a happy life, full of adventures.

Like every beginner, I began collecting everything all at once: space flights, sports, art, fauna, flora, and so on. I was shamelessly taken in right and left. For a period of time, I was a chump, a simpleton, easily misled into buying a less-than-perfect stamp.

My first serious trial was a small collection of English stamps from a man named Kahn. He slipped in a number of fakes, but he forced me to sit down and think seriously. I joined a collector's club, bought catalogues, and started to read special literature, listening and leaning toward serious collectors and gaining experience at the same time. I had time and, most importantly, money to spend on stamps. I didn't have a big salary, but I had a decent private practice. I should emphasize the latter: sellers knew that I could pay well for a good stamp; thus, it was harder to saddle me with a fake.

After some time I gained a reputation as a serious collector. I was often approached for consultation and advice. I studied all the catalogues—Yvert, Michel, Gibbons, Scott, Zumstein—and after some time, I made enough good purchases and exchanges to get back the money I had lost on fakes. I never perceived philately as an investment. I felt that it was a passion, one of the strongest and happiest in my life. For a rare stamp, I was ready to not just rush across the city, but to travel a thousand miles from Moscow.

I had a neighbor, Mikhail Volpianski, a ninety-two-year-old, well-respected lawyer, who had been through a lot in his life. He knew the last czar's family and both before and after the war, he managed the famous Yeliseyev Food Store. He began to collect stamps before the Revolution and was seen as a patriarch by collectors. He made a living estimating collections and selling them for owners. I, too, bought tons of stamps from Volpianski, including lots of junk and fakes, but in time, I acquired knowledge and experience and, with his help, purchased a number of valuable stamps.

Volpianski consulted Marshal Climent Voroshilov, who also had a weakness for stamps. The latter's opportunities were limitless, especially at the end of the war, when the Soviets took everything they could out of Germany and other countries, including stamps.

Academician Lifshitz and Slutsker told me how old collectors would meet trains arriving with soldiers returning from Germany and hold up cardboard signs that read, "Will Buy Stamps and Pay a Lot." They bought suitcases full of stamps for peanuts.

I also heard that appropriated stamps had been taken to a special warehouse. It is easy to imagine a warehouse stacked with sacks and suitcases, all filled with rare stamps. One can only imagine the treasures! Lifshitz was one of those few lucky people with a pass to this warehouse. He was left alone with treasures and could dig through the stamps for any amount of time, choose any he liked, and pay whatever price for them he felt appropriate.

This obsessed collector had a brother, a theoretical physicist, by the name of Evgeny Lifshitz. He was famous for his work in iron magnetism, phase transitions, cosmology, et cetera. With Landau, he authored a classical course on theoretical physics. Ilya Lifshitz (the collector) founded his own school for solid-body physics and created an electronic theory of metals. It is safe to say that he was as much of a legend among collectors as he was among physicists. His knowledge of stamps was immense and matched his collection, the greatest one in the USSR. It was a veritable stamp museum. He never sold a stamp in his life, even the simplest one, even if he had a hundred identical ones. If he had a chance, he would buy #101.

Late in life, Lifshitz became very ill, suffering from angina, and money was no longer as plentiful. Yet he continued to spend all his money on stamps, paying no attention to his wife's complaints and ignoring their living expenses. I remember suggesting, that he sell some of his treasure, to help his financial situation. We discussed this for some time and he even estimated how much his collection of stamps from Austria, which was a small part of his total collection, might fetch, but then, immediately he got a hold of himself: "I will not sell anything—don't even dream of it! I was just speculating."

"What about Holland?" I asked, since I was into Holland stamps at the time.

"Are you out of your mind?" he demanded. "Selling Holland means taking apart Benelux! Without Benelux, the European southern flank will be open!"

Then I realized that speaking with Lifshitz about selling stamps was absolutely useless. They were to him like an army to a general.

I also recall a collector named Steiner. He collected only German stamps. He knew that country very well and had a whole library of publications on German stamps. He lived in poverty, spending every penny on stamps. He even divorced his wife, in order not to spend money on his family. Despite his huge effort, he was still missing a few German stamps that Lifshitz had, even in several copies. Steiner begged the scholar to sell those duplicates, assuring him that without them, he absolutely could not live. But Lifshitz stood fast.

"Listen, I have only five copies, and I want more. If I find them, I will definitely buy them," Lifshitz said.

Without these stamps, Steiner could not put his fabulous collection on display. The poor man was so upset that he had a heart attack that took his life. Now, one can imagine what unbearable suffering Lifshitz experienced when he saw a stamp that he did not yet have, and he felt the same because of the death of Steiner.

Few can brag about being able to impress Lifshitz. I am proud to say that I did it a few times.

The Golden Age for Soviet collectors fell in the forties and the fifties, when collections were created on the cheap, and rare stamps were sold and resold. I came too late and caught but a tiny feather from the heavenly bird of luck. By the time I became serious about stamps, the market was solidly tied up by expert dealers, who knew perfectly well all the serious collectors in all the cities; they knew who collected what, who died where, and who inherited the collection. A great deal of stamps left for the West, but many stayed behind, too. Thus, patience was occasionally rewarded, with a few moments of fabulous luck. Like every collector, I know a few such stories.

After the war there lived a famous gynecologist in Leningrad, who was a passionate collector. All of his stamps were in ideal condition; he did not buy any other kind. Through a mishap related to his medical profession, he was imprisoned. Being an exceptionally well-prepared man he hid his collection away in advance. Thus, though his property was confiscated, he kept his collection. After his release he went back to it, adding new stamps.

As he and his wife had no children, they adopted a Georgian boy from an orphanage. The boy grew up to be an alcoholic. When the doctor died, the widow locked up the collection. Yet, as the saying goes, a lock works against honest people but never against thieves. A friend of the doctor named Vasiliev visited Leningrad from Tallinn from time to time and got the adopted son drunk. Somehow, the latter gained access to the collection. The results of this trade for "firewater" didn't surface for a long time.

I was once introduced to the gynecologist's widow. She allowed me to see the collection but warned me, "I do not want to sell the stamps, but I will show you, since you are also a doctor. Call me in about a week."

When I called, the son picked up the phone and informed me that his mother had died. Not believing my ears, I flew to Leningrad to visit him. From their neighbors, I found out that after speaking with me, the widow had opened the closet where she kept her husband's collection and discovered only empty shelves. The alcoholic son had sold everything. She had a heart attack and passed away.

I knew Vasiliev. I saw him on my visits to Tallinn and had bought a couple of things from him. I once bought a collection of Gibraltar stamps that had previously belonged to the gynecologist. Vasiliev kept the best copies for himself and offered me the rest. On one occasion, he came to Moscow and visited me to show me the famous "Napoleon," a rare French stamp in ideal condition. This stamp is so valuable it deserves to be described in detail.

The first postage stamp was issued on May 6, 1840, in England. On that exact day, Rowland Hill, the inventor of the stamp, wrote in his journal: "I woke up at eight in the morning. For the first time stamps are released today to the London population, terrible commotion at the main post office!"

The first French stamps appeared nine years later. The Napoleon stamp shown to me was very well preserved. I was completely captivated and hypnotized by it. I was ready to purchase it for any amount of money, exchange it for anything. Vasiliev said that he had just acquired this stamp and was not ready to sell it. He had brought it to show Lifshitz; the stamp was in such good condition that its lucky owner Vasiliev questioned its authenticity. I had similar doubts. Later, I found out that Vasiliev had bought this stamp practically for pennies from the late gynecologist's son.

A couple of years later, Vasiliev died. His widow did not know much about stamps, and his chauffer son knew even less, being better acquainted with alcohol. A friend of Vasiliev's, an administrator at Tallinn Philharmonic named Gunnar Pyluas, knew the worth of stamps very well. Pyluas collected Finland. When he came to Moscow, he gave me a call and offered to meet with me.

"I want to show you something and trade you something from your Finland collection," he said.

This was a strange offer. Essentially, I never sold or traded; I only bought. When we got together, Pyluas opened his album, and I almost fainted. Napoleon! I had no doubt in my mind that he had bought this stamp from Vasiliev's alcoholic son. I summoned all my strength to conceal my excitement and continued looking through other stamps in the collection. Then I showed my Finland collection. He chose what interested him, and I chose what I needed. If he had asked me to trade my entire Finland collection for the Napoleon, I would have, without a second thought. But it was very important not to lose face. I can't tell you how many times other collectors beat me in this game, but sometimes I got lucky too. In passion and excitement, this game is without peer! That time, I managed to win the Napoleon.

As soon as Pyluas left, I called Lifshitz—I couldn't wait to have my trophy authenticated. It turned out that Ilya was resting at the sanitarium. My wife and I got into the car and rushed to see him. Needless to say, my visit excited the scholar. He was a big star in theoretical physics, so famous that scholars from other cities had to wait for months to see him. But when I called to tell him that I got a new little something, he demanded that I immediately come and see him. He was very impatient to see the "new little something" because he knew that the item was probably of great interest. When I showed him the Napoleon, his hands trembled, and he turned pale as he feverishly swallowed a nitro.

"Where did you f-find t-this s-stamp?" he stuttered with excitement.

Of course, collectors never reveal how they got hold of a rare item, so I remained silent. Ilya, however, insisted, demanded, and pleaded. Finally, I could no longer resist the onslaught and told him.

"Wonderful! Fantastic!" he exclaimed. "You know I don't trade, but for this one, anything you want!"

The trade was out of the question, since I would have given up everything I owned for the Napoleon. The most important thing was that Lifshitz was almost completely sure the stamp was genuine. Still, he told me to look it up in special literature, check certain details and strokes, and inform him of the results. I didn't even get home before Lifshitz called me and asked "Well?" I promised to return his phone call as soon as I had checked the stamp. The stamp turned out to be genuine!

Another story deals with General Kolosovski's collection It was said to be more a warehouse than a collection.

The general was just in the thirties. He must have felt what was coming and kept his collection elsewhere. When collector died, the collection remained. It was kept by his daughter, a chemistry professor. She had offers from potential buyers, but she did not want to hear about it. The most persistent one was the famous collector Blechman, an engineer. He was an official Soviet stamp expert, a winner of many prizes at international stamp exhibitions, and he had known the general. Moreover, his first wife had gone to school with Kolosovski's daughter. Thus, he had a better chance than anyone else to get hold of the collection. To that end, he created something like a pool or a stockholding entity, since no individual collector would have the means to buy Kolosovski's treasure. Of course all the discussions, even the daughter's address, were kept in secret.

My Stamp Collection

There were about fifty women living in Moscow with the same last name as the general's daughter, Kolosovskaya. I called almost all of them before I finally stumbled upon the sought-after address. To make sure, I checked with the super—yes, it was the right Kolosovskaya. As I mentioned, by then she had become a professor who was married to another professor, and they had an adult daughter.

I kept postponing the visit for several weeks. Finally, I overcame my shyness. Her husband opened the door. Once he found out I was a doctor, he didn't exactly rejoice but at least treated me without enmity. He warned me that his wife "showed the door" to anyone who mentioned the collection.

"Consider yourself lucky that my spouse isn't home," he told me. "Call me in a week. I will try to find out—carefully—how my wife feels about your offer. You are not just any middleman, but a serious person—a doctor. You know, our daughter is grown up and needs her own apartment."

I was restless for the next week. Finally, I called and spoke with the husband. "I came to see you a week ago."

"I remember," he said. "I am sorry to disappoint you, but my spouse has already sold the collection."

"How come?" I exclaimed.

"She went to see an old stamp collector, an old friend of her father's, who persuaded her not to deal with strangers. He had a trustworthy friend who wanted to pay a good price for the collection. He swore on the general's memory that everything would work out fine."

The collector sent over his friend Aleksei, whom I knew very well. He did not have money. However, he did know Stishinski, son-in-law of a church bishop. Stishinski used to come to collectors' meetings with wads of money and buy any collection for any amount. Eventually, he got hold of General Kolosovski's collection.

Another situation, however, turned out in my favor. There was a well-known collector in Kharkov named Aleksandr Ilievich, who was a professor of radiology. When we met, he was getting on in years. He once showed me an orange "five-pounder" issued on March 21, 1882—an enormous English stamp with Queen Victoria's profile inside a medallion. This was a very rare stamp. I had a cancelled one, but I had only ever seen a clean one in Lifshitz's collection. I asked Ilievich if he were willing to sell the orange stamp.

"It is the same thing as selling your life," he said as he tenderly eyed his favorite, encased in glass, like a painting.

Whenever I came to Kharkov, I always visited Ilievich to admire his stamp. A few years passed. His daughter got a divorce and was left alone with her little daughter. The girl was growing fast. Once, the daughter confessed that they had fallen on hard times, but her father would not hear of selling the stamp. Yet eventually his daughter's constant complaints and her persistence took the edge off his determination. At the same time, the professor was old and helpless. When his daughter visited him, he would start trembling, scared she would bring up the subject of selling the stamp. A couple of times she almost talked him into it, but at the last minute, he would always find the strength to refuse.

Meanwhile, the granddaughter had grown up and was getting ready to go to Moscow University. Her mother tried to use this moment to launch the decisive offensive.

"We might sell you the stamp," she said to me on the phone. "You can come."

I left for Kharkov that day, and arrived at Ilievich's the following morning. His daughter warned me, "It will not be easy. Let's try together."

The professor sat without taking his eyes off of his favorite stamp, as if saying good-bye.

"I understand," his daughter said. "But you are not doing it for yourself or for me, but for your only granddaughter. Do you want her to enter the university? Yes or no?"

The old man would not move. There was so much suffering in his face that I could not take it.

"I'm sorry," I said to his daughter. "I cannot go on. I'd love to have the stamp, but you see that your father cannot part from it."

"Please wait in the other room," she told me.

In a moment, she came back. "We have already made a decision. Even before you came, he agreed to sell it to a professor from Leningrad for two and a half thousand. But if you agree to pay three thousand, you can have it."

My hands trembled as I counted out the money. The teary-eyed professor took the stamp out of the frame with his pincers and set it on the table. I was in a hurry to leave before he changed his mind. But he begged me to hear the history of the stamp.

My Stamp Collection

He had bought the orange stamp in a Torgsin store in the twenties—a special store where you could buy things for foreign currency. During the war, his collection stayed back in Kharkov and was lost. Only this one stamp, wrapped in plastic, spent the war with him in the pocket of his military uniform.

Back in Moscow, after admiring my new acquisition, I went to see Lifshitz and asked to see his amazing English collection. As I flipped through the album, I noticed a "five-pounder."

"How long have you had it?" I asked him.

"Very long," he said, adding that he had bought it from an old collector whom I didn't know.

"Who knows ..." I said vaguely. "By the way, I recently got hold of one just like this."

Lifshitz froze. "I know that you do not reveal your secrets, but please make an exception for this stamp. I give you my word, I will not approach this person or say anything about him. I am dying to know where you dug up this stamp."

"If you promise to keep it a secret, then I'll tell you. I bought it from Ilievich."

Lifshitz collapsed, grabbing his chest. He seemed to be having a heart attack. "This cannot be! You are joking with me! Or someone sold you a fake!" he said.

"Why can't it be?"

"Ten years ago I bought this stamp from Ilievich," he said.

Now my knees began to get weak. We argued; each was positive that the other one had a fake. Lifshitz was from Kharkov, too, where he graduated from university and was in charge of a department at Kharkov Tech. He knew Ilievich very well, so he called the professor and demanded an explanation.

"Ilya," Ilievich said, "please stop worrying and getting angry. Way back, I bought two identical stamps. I sold you one, and the other to another person, not too long ago."

The scholar nearly had another heart attack and swallowed about ten pills of nitroglycerin, but nothing would calm him down.

"Congratulations! Your copy is the better one!" he said. "When Ilievich sold me the duplicate ten years ago, he had kept the better of the two for himself."

It turned out to be true. Lifshitz's stamp carried a black fleck, but mine was spotless!

Here is the story of the famous "inverted Levanevski," a story shrouded in secrets. On February 13, 1943, the Soviet boat Cheluskin, sent from Murmansk to Vladivostok, was crushed by an iceberg in the Chukotsk Sea. A team of airmen—Lapidevski, Levanevski Slepaev, Doronhin, Vodopianov, Molokov, and Kamanin—rescued the sailors from a drifting floe and was later awarded the title of Hero of the Soviet Union. On January 25, 1935, a series of ten stamps was released to commemorate the rescue.

Some time later, Sigizmund Levanevski dared to make a nonstop flight over the North Pole. To commemorate this flight, scheduled on August 3, 1935, a red imprint reading "Moscow to San Francisco over the North Pole 1935" was made on forty thousand stamps with Levanevski's picture. The imprint was made in a rush, and a number of copies contained a typo: a lower-case "f" in the word "San Fransisco," instead of the capital "F." Eight thousand stamps had a typo. This made them rare. Rarer still became the stamps with the imprints made upside-down—there are rumored to be only a hundred of them. Yet the rarest are the ones that have both a typo and an upside-down imprint. The Soviet government gave one as a gift to Franklin Roosevelt in 1945. After the death of the American president, his collection was sold, and the upside-down Levanevski was said to have ended up in a private collection belonging to the Queen of England.

For a long time, the printing mishap was explained as an accident, a pane placed upside-down on the printing press. Yet in that case the upside-down text would have appeared on the dark side of the stamp, rather than where it did, on the light side. Today, experts are convinced that the mistake was deliberate: the list (some sort of press) was not placed upside down, but the lines were shifted. Some point the finger at Henrich Yagoda, the interior minister, who ordered that a rare stamp be created.

I had been looking for this stamp for a very long time and lost hope of ever finding it. Once, I happened to meet a collector named Insarov, whose wife's sister was married to Tikhon Khrennikov, an iconic Soviet composer. Insarov was very ill and, knowing he had little time left, decided to sell his collection. It wasn't of much interest to a serious collector—at least, I wasn't too impressed. I was ready to take a pass when Insarov opened a small album—and I beheld an inverted Levanevski.

I could not believe my eyes. Once I calmed down, I thought it could have been a fake. This type of expensive item should not find its way into a small collection.

Insarov said he had bought it from someone who worked as a censor at the stamp-printing shop or was connected in some way to stamp printing. I did not believe a word of this nonsense. Collectors make up better stories than novelists. I told Insarov I was ready to buy the stamp, provided it was authentic. He assured me it was. My instinct also told me that the stamp was real. I threw caution to the wind and rushed home to get the money. I left as the owner of the well-known stamp.

But the worm of doubt still gnawed at me. I took the stamp and went to see Blechman, who knew everything there is to know about Soviet stamps. He also had his very own inverted stamp, though his was cancelled. Needless to say, Blechman was very shrewd.

He inspected the stamp for a long time, first with a magnifying glass and then by raising it to the light. Finally, with a sad smile, he said, "A fake, yet a remarkable one. What did you pay for it?"

"Three thousand seven hundred," I replied.

"I have a business proposition for you. I want to have a clean inverted stamp, even if it is a fake, just to fill in the gap in my collection. At least you can make up a part of your expenses. I'll pay you a thousand."

By then I knew the psychology of collectors quite well—all their tricks and ruses. If Blechman offered a thousand, it was the best guarantee that the stamp was real; he wouldn't pay a ruble for a fake. Yet I still had doubts—after all, I had paid some really insane money.

I ran to Insarov, wishing to return the stamp, but he refused, fearing I had made a switch. How grateful I am now for his refusal! Yet once again, at the time, his refusal made me suspect the stamp was a fake, and I decided to check it again.

Soon, I heard of a big stamp show in Kiev coming up. One of the guests was to be a Monsieur Lipshotz, a great expert on Soviet postal rarities and a member of the French Philately Academy. I went to Kiev specifically to show him the inverted stamp. He already had it in his collection. He carefully inspected my stamp. "I think it's authentic. Where did you get it?"

Even he, a great collector, was curious where I got my stamp. I could now begin to relax. However, my joy was still marred by the shadow of doubt. Judge for yourselves: two of the world's greatest experts gave opposing conclusions. Blechman said "fake," while Lipshotz said "authentic." I strove for final, absolute clarity. This could be done only in Leningrad, in the Popov Central Communications Museum, which held the official State Stamp Collection.

I went to the government with the request to allow me to visit the museum, in order to compare my stamp with the official copy of the inverted stamp. My request was finally granted. When I finally arrived at the museum in Leningrad, I told the director of my problem. One could see the stamps only in the presence of several colleagues. The most important moment finally arrived; they brought out the album and displayed the famous stamp as I took mine out. We compared the two copies thoroughly and didn't find a single difference. The inverted stamp turned out to be authentic. Now I could feel lucky!

At a collector's meeting in Leningrad, I met Boris Kudryavtsev, one of the oldest collectors in the city. He was a chemistry professor, long retired, but he continued to buy stamps. He never sold or traded— allegedly, no one had ever been able to get a stamp out of him. No one could even see his collection. Someone once tried to rob him, but the screams alerted his neighbors, and the attempt failed. After this, the professor was even more cautious.

Kudryavtsev collected everything, so when he was offered a stamp, he could never remember if he had it or not. He would turn away from the visitors, find the right album, open it, and stick his head in between the pages. Assured that he had the stamp, he would put the album in the closet, lock it, place the key in his pocket, and rush the guest out. "I already have that stamp. Good-bye!"

He recorded every guest in a specific notebook: "Such-and-such came by on this date and offered this item for this price," and so forth.

This aged, sickly person, full of oddities and quirks, treated me well. Perhaps my profession inspired trust. I explained to him that senility was a serious illness, especially at his age. If, God forbid, something happened to him, his collection would go for pennies.

My Stamp Collection

This is what usually happened to most collections. Blechman's collection was a vivid example. He died of a heart attack in his country home. His widow panicked and feared that she would be robbed. She did not feel safe behind her steel door, tricky locks and alarms.

Once, she spent the night at her relative's—and on that very same night her apartment was robbed. The entire classic collection of the late Blechman was stolen. This detective story remained unsolved. The most experienced detectives were assigned to the case, but they came up with nothing. Some angry people gossiped that the widow faked the robbery, so she would not have to share the proceeds with her husband's first wife and son.

After some time Kudryavtsev began to trust me so much that he showed me one of his collections. It was absolutely unbelievable. He then began to sell me stamps, though his negotiating style was quite unique. He would ask, say, fifty rubles for a stamp that I knew to be worth only ten. I would show him the catalogue and he would say, "If you don't want it for fifty, then pay sixty!"

Further argument led to a higher price, so bargaining was useless. You could either agree or leave. Sometimes, he offered a collection for a thousand and then added, "A thousand plus ten rubles for the tomatoes."

If the buyer wondered why he should pay an extra ten rubles for "tomatoes," Kudryavtsev would immediately double the price. Yet I am grateful for the rare stamps that I had the opportunity to buy from him.

The collecting passion could be so intense as to make stamps more important to a person than his life. I already mentioned Steiner, who could not live with the knowledge that his collection lacked the German stamps that Lifshitz had in several duplicates. Gunnar Pyluas from Tallinn, from whom I bought the Napoleon, was also such a collector.

At first, Pyluas sold only Soviet stamps, but then he became interested in and began to collect Finnish stamps. He once bragged to me that he had the most amazing collections of stamps from Sweden, Switzerland, and Germany. Something in his story sounded suspicious. I had heard of a collection described similarly before. My disbelief was so strong that he confessed that the collection did not yet belong to him but would soon, for he had a firm promise.

Longevity and Health

I gradually discovered that the collection in question belonged to a famous collector, Maazik from Tartu. With a great deal of effort, I located his widow, a lovely old lady who lived with her daughter and son-in-law. It turned out that the widow had no intention of selling her late husband's collection.

Some time later, I found myself in Tartu and stopped by the widow's house. The old lady loved the suit I was wearing. After complimenting me on my taste, she blurted out unexpectedly, "This time you will not leave empty-handed." She sold me a valuable part of the collection with many excellent stamps. Now I was finally sure that Pyluas had been describing Maazik's collection.

Tartu is a small town. A local collector recognized me, and word reached Pyluas. He called and told me he knew what had happened, and of course he was upset, since I had clearly bought the best of the collection. Some time later he had a stroke and became blind in one eye. This is how much he was shaken.

My path was not all roses, however, and I got bloodied by quite a few thorns. One of the most unpleasant incidents happened about two years before I left the Soviet Union, in 1988. I neglected to lock my car (having stepped out for a couple of minutes to go to the pharmacy), and someone stole my briefcase, which contained a large collection of German and Spanish stamps. I was horrified. Had someone tailed me specifically to steal my album? Or did they steal the briefcase without knowing what was inside?

I immediately reported the theft to the police headquarters in Petrovka, but I got no results. Then I undertook a personal investigation. Believe it or not, I succeeded. I was not as brilliant as Maigret or Colombo, but better than the Moscow police, who turned out to be a bunch of incompetents. I set up an entire system of traps, ambushes, and bait among Moscow collectors, and I discovered the thief's trail. I did not get everything back, though I was able to buy back most of the stamps stolen.

To conclude this chapter, I want to share some observations on stamp collectors' mentality. This is their prominent trait: they are happy to get hold of a rare stamp, but they are twice as happy if they do so for pennies. The greater the difference between a stamp's true value and the price paid, the greater the happiness. Perhaps this is common to all collectors.

My Stamp Collection

Slutzker, the patriarch of Russian philatelists, who inspired me to collect stamps, was happy like a child when he was able to dupe someone. He was one of those people who met soldiers coming back from Germany and bought stamps for a pittance. I think he remembered what price he paid for every stamp in his enormous collection.

"This one I bought for three rubles, though its real price was four hundred. This one I bought for three rubles, too," he told me.

"What if it costs a thousand?" I asked naïvely.

"That's none of my business. A man asks for three rubles, he gets it. I don't have to share my knowledge with him."

I witnessed one incident myself. Someone was selling his stamps, fifty kopecks each. Lifshitz bought one but took it out of the album in such a hurry that a tiny corner ripped off. He seemed to ignore it, though such negligence was not like him. I asked him about it.

"I would have given myself away! If a buyer picks up the ripped piece, that means the stamp is valuable, and the owner will demand more money for it."

The stamp that the scholar had bought for fifty cents turned out to be very valuable: Hungarian stamp #1! This stamp was very rare and in very good condition, bringing the price up to several hundred Euros.

I had my fair share of getting duped. I once bought a huge collection of English stamps that contained many envelopes. I didn't collect them at that time and sold them to Victor Rosov, the playwright, who collected both stamps and envelopes. We traded stamps for envelopes, pricing each envelope at two, three, or four rubles.

About a month later, I decided that a serious collection must include envelopes, which would serve as decoration. I ran to Rosov to buy back my envelopes.

"No problem!" he agreed. "I have your envelopes, but the price has changed."

"What price are they?" I asked in surprise.

"They now run from fifty to a hundred rubles per envelope," he replied.

I was in shock. "Victor," I said, "Have a heart. Why are you asking so much for my very own envelopes?"

He told me he had found a catalogue of English envelopes with prices, so "Why should I sell my envelopes that cheap?"

Longevity and Health

Serious collecting is a sophisticated business that demands a lot of special knowledge, gained over years. Even the most knowledgeable collectors sometimes make mistakes.

I once bought a Chinese stamp with a surcharge from Rosov, for one thousand rubles. After a couple of years in Holland, I sold it for eight hundred German marks. Six months later, this stamp was sold at an auction for forty-two thousand German marks. Who knew that a few Chinese collectors would compete for it all at once and its price would skyrocket?

I do not try to appear an angel amid my colleagues. There were plenty of times when I bought a stamp for a price much lower than its worth. I know perfectly well all the catalogues, and when it suited me, I would show the buyer a price from one catalogue, though other catalogues cited much higher prices. Stamp prices may differ tens of times, depending on the type of catalogue.

On the other hand, I had to overpay sometimes, too, though I never regretted it. I knew that collectors gossiped about me, but I understood that this was envy. As a rule, the best stamps fell into my hands. This happened thanks to my knowledge of stamps and the availability of funds. Most collectors had neither.

I could recall many more other stories related to stamps,[1] but this is all in the past. Having seen what befalls some of the biggest collections after the death of their owners—squabbles between heirs, theft, sales for pennies—I promised myself to quit collecting at age sixty-five and then sell them.

When at first I began to sell my stamps, in Holland, it seemed as if I were selling a part of my soul. Yet after some time, I grew cool toward philately, and medicine filled my life once again. From many years of contact with philatelists and the biggest philatelic companies, I came to the definite conclusion that they were not honest and that they had tried to buy my stamps as cheaply as possible, always causing me to lose a lot of money.

---

[1] I wrote a book called *Passion and Tragedy. A Philatelist's Confession*, which deals with crimes and dishonest acts committed by both private collectors and large firms.

My Stamp Collection

Myself examining stamps.

Chapter 8

The Dutch Maze

Before 1988, I had never been out of the country, not even to Bulgaria as an ordinary tourist. Sometimes I asked for permission to go abroad, but the answer was: "Not in the State Interest." The real problem was that I was working at so-called "special hospitals," where top Party members were treated. Any information about the leaders' health had always been a State secret.

In May 1988, my wife and I flew to Holland, on an invitation from our daughter. A few years earlier, she had married a Dutch doctor and gone to live in Amsterdam.

Our plane landed at Schiphol airport, one of the busiest in Europe. Within the first couple of minutes, I was amazed by the smoothness, order, and simplicity of all the formalities that usually scare Soviet citizens so much. I was awed by a payphone; it was hard to believe that you could call Russia, America, or Africa from any payphone. Equally shocked was the woman at the post office, when I inquired as to how I could call Moscow.

"Use the phone," she said, looking at me with surprise.

"What telephone?" I asked.

She was even more shocked.

"Any phone! There is a phone outside that you can use."

Since I did not speak Dutch and very little English, I felt uncomfortable. I sometimes got into some funny situations. A Dutch friend once recommended to me that I visit the Fon Hoch art museum.

"Fon Hoch? I never heard of him," I said.

"What do you mean? This is one of the greatest Dutch artists! Well," he smiled, "it's okay. You are from Russia, aren't you?"

Later I learned he spoke of *Van Gogh* but in Holland, they pronounced his name *Fon Hoch*. There were other similar confusions, until I began to catch on to Dutch pronunciation.

Unfortunately, the euphoria of my first couple of days soon gave way to disappointment. As, often happens, it came down to a trifle: a cup of coffee. My wife and I stepped into a café, and being shy about my pronunciation, I ordered a cup of tea and a small cup of coffee. It cost us five guldens, or two dollars. In Moscow, this cost only sixteen kopeks. A bus fare also cost five guldens. In short, the two thousand guldens of travel money that had made me feel wealthy turned out to be a relatively small sum. It appeared that only with a large amount of guldens, marks, or dollars in one's wallet could you enjoy the advantages of Western life.

Still, I was making plans, hoping to continue my research on delivering drugs into the brain using iontophoresis and my research on the autonomous nervous system, using a system I called electroautonomography (EAG). One should bear in mind that I knew no one and nothing in Holland, and no one could give me a hint about where to begin. I had to act blindly, thus wasting a lot of time.

One day, I stepped into a clinic that turned out to be a hospital for chronically ill patients. I introduced myself as a doctor from Moscow and said that I had invented new methods of researching the nervous system, had a Soviet patent, and wanted to offer my research to Holland. The doctor with whom I spoke gave me a list of clinics, along with addresses and last names of a couple of doctors who were doing research in electrical physiology.

The first name on the list was Professor De Wisser from the Amsterdam Medical Center (AMC). I called him and asked to meet with him. He was interested in my story and was extremely surprised. "The Soviet Union is such a huge, powerful country! Why are you unable to finish your research there? To me it sounds highly promising."

De Wisser promised to call, but day after day passed without a phone call. Before I left, I called him once more to say good-bye, and he unexpectedly offered to meet me. When I came to see him, he introduced me to a Professor from the Academic Medical Centrum, whom he described as an expert in neurology. (this professor caused me great difficulties, which I will explain later.)

The Dutch Maze

I happily returned to Moscow knowing that I had made a contact in Holland. When we visited our daughter again, in April of 1989, many things were already familiar to me. My son-in-law, who was familiar with my work, suggested that I speak with an ordinary advisor Tos Schunovale, who had worked in Canada for many years and recently returned to his homeland. My son-in-law had also spoken with him and thought his advice would be helpful in realizing my plans. Aware that Schunovale had made enough money in Canada to retire and build a house in Holland, I decided that it could serve as evidence of his wonderful business ability and I took my son-in-law's advice.

As soon as we met, Schunovale suggested we drop formalities. "Just call me Tos." He was very interested in my methods and wanted to help me continue my research and construct the devices I had invented. He called me on a daily basis for a month, always asking me new questions, yet we never moved far from square one.

Finally, at the end of my visit, Schunovale asked me to lunch.

"I have a serious proposition for you," he said.

The "serious proposition" turned out to be a contract permitting Schunovale to keep looking for investors in my absence. I got a little excited at first, when I realized that the entire contract was written in Dutch, I quickly lost my desire to sign. This was my first experience in communicating with Western businessmen.

I then met Dr. W. J. Maylor, an anesthesiologist from Groningen, a city in the north of Holland. He offered to let me give a lecture in his department. After I lectured, two professors, a neurosurgeon and a neurologist, showed interest in joining forces to conduct research. About that time, I also met Tanstijn, a representative of a big Los Angeles company selling medical equipment in Holland. I introduced Tos and Tanstijn to Dr. Maylor, and had now assembled a small group of people interested in the results of my work.

Dr. Maylor offered to demonstrate my device at the RUG university of Groningen in the North of Holland. A close friend of his named De Boer also attended. This generated the idea that we could collaborate. De Boer asked me to bring my documents, a brief autobiography, a copy of my diploma, and a number of other papers necessary for insurance. Enthused about his offer, I brought him the papers and was shocked at his reaction. He did not display any enthusiasm or interest whatsoever. He picked up the papers reluctantly and said he would try to do something but couldn't make any promises. The situation had obviously changed. But what could have happened? I felt that something was working against me. Later on, when Maylor started acting suspiciously, I realized that it was he who had persuaded De Boer to withhold assistance.

About twenty to twenty-five professors, specialists in electrical physiology and neurology, attended my second lecture at the university. The discussion that followed was very lively. I was asked to demonstrate in the laboratory the processes that I explained at the lecture, but I remembered the warning of an expert at the patent bureau: "Mr. Lerner, be careful until your research is published. Scholarly theft is very common among doctors."

So why exactly do scholars, doctors, and businessmen show such interest in my research, investing in it, plotting around it, and stooping to flat-out lies in drafting agreements, contracts, and other papers? In order to understand this, it is necessary to explain what electroautonomography is and where it comes from.

Electroautonomography (EAG) is a method of researching the autonomous nervous system (ANS) by registering skin potentials (electrical skin activity). It is based on the same principle as electrocardiography (EKG) and electroencephalography (EEG), widely used in diagnosis and research. The founder of electrocardiography and the inventor of the device for registering electric changes in the heart was a Dutch physiologist, Willem Einthoven, who was awarded the Nobel Prize in 1924, for the discovery of the electrocardiogram mechanism. Electroencephalography is the graphic registration of bioelectrical activity in the brain, and the device was invented by the German psychiatrist Hans Berger, in 1931.

At the time, these methods brought about a revolution in medicine. Until recently, there has been no method of directly studying the autonomous nervous system, though it regulates blood circulation, digestion, and breathing, and its disturbances may lead to hypertension, strokes, heart attacks, digestive disease, and much more. In 1999 and the years that followed, I have worked with several universities, including Vanderbilt University, in Nashville, Tennessee, in the United States. After using my electroautonomograph in scientific and practical work for seven years, that university sent me a letter in 2006 confirming its usefulness (I have included this letter in the appendix).

To receive information on ANS, you must complete dozens to hundreds of tests, which makes the observation long and expensive. My proposed method of electroautonomography (EAG) allows the recording of the potentials of areas of skin with special electrodes and therefore the collection of necessary information in just fifteen to twenty minutes without resulting in high expense.

Although studies of diagnosis by skin potentials that measure the electrical fields of the skin started over a hundred years ago—and by my count, over ten thousand scholarly works have been published on the subject—the colossal collected material is still not used in practical methodology. This is exactly why my EAG method immediately drew doctors' interest. They especially liked that it could not only diagnose but also predict a number of diseases. On the basis of EAG data, one can predict a definite potential for a serious disease, like high blood pressure in an infant. This opens the door to developing better strategies for prevention.

A group of American doctors arrived at the conclusion that EAG can be applied in practically every area of medicine. That means that EAG will become as common in clinics and practicing doctors' offices as a cardiograph and tonometer. Every person will be able to have the condition of his or her own autonomous nervous system examined.

As for businessmen, my method made them think of a new gold rush. According to preliminary calculations, if one in ten U.S. doctors purchased an EAG, the manufacturers would make many millions of dollars (to say nothing of Europe). This was obvious to anyone who acquainted himself with my work. Once I realized the scale of my project, I became aware of the complexity of the situation I was in.

Let me return, however, to my presentation at the university where Dr. Maylor worked. The doctors contacted the queen's commissioner, and we went to see him. The queen's commissioner is omnipotent in his region; he can issue permits for work and residence. When the neurology professor and I came to see him, he already knew about me.

"Dr. Lerner," he said, "in this room, you can say anything without fear. Not a single word will be used against you. Honestly, why did you leave the Soviet Union, and what do you want to accomplish while here in Holland?"

I sensed his good disposition and told him a little about myself: where I was from, where I had worked, what problems I had had. The commissioner listened attentively and occasionally nodded in agreement. When I finished, he said, "Dr. Lerner, we will give you a residence permit, give you a job, and you'll be all set. Everything will be fine."

Just then the professor pulled me aside and said, "Dr. Lerner, please wait in the hall; we need to discuss something with the commissioner."

He came out of the commissioner's office about twenty minutes later. I thought I would be invited back inside, but that was not the case.

"Let's go," the professor said. "Everything is settled. I'll tell you on the way." He went on to tell me that while I could be granted a work permit for six months, my wife would have to go back to Russia.

I was stunned. The commissioner had mentioned neither a term nor my wife's departure. It was clear that while I was in the waiting room, the doctor had said a couple of things that radically changed the commissioner's decision. I could not go back to the commissioner, because you must be invited to enter his office, and so I was forced to refuse this offer.

A number of times, I was offered a contract with Schunovale and Tonstijn. Even Schunovale's daughter met with me twice, trying to convince me to sign up with her father. The bottom line was that, in any event, I would give them half of my income for my entire life, no matter if they were in Europe, Russia, or in any other country. God must have stopped my hand from signing.

My next visit to the Land of Tulips was in the fall of 1989. Tonstijn invited me over this time, but my relations with him and Schunovale were getting worse every day. I was tired of being a bear whose hide these gentlemen were always ready to split open, and I said as much.

Sensing that the breakup was inevitable, Tonstijn did something unexpected. He suddenly said with a hurt look, "Whether you want to sign our contract or not is your business. At least compensate my expenses and return the ten thousand guldens that I spent on you!"

"Ten thousand guldens! Where did that figure come from?"

"Your hotel rooms and your dinners in restaurants!"

"You did not pay for my hotels. As for dinners, they never went beyond a cup of coffee or tea, and, as a rule, I paid for both of us!"

The Dutch Maze

This was obvious extortion, though it went beyond that. Now Tonstijn went for blackmail. He doubled the sum in question, and threatened to demand it through the Soviet Embassy in The Hague. Twenty thousand guldens! I had never in my life held that amount of money in my hands, although my stamp collection was indeed also very valuable. Even if I had such a sum, I would never pay, since I knew perfectly well that they were bluffing. My daughter met with Schunovale and shamed him for his improper behavior. Nonetheless, in order to get rid of him, I offered him nine hundred guldens. He took the money and signed off that he had no further claims against me.

The conflict appeared to be settled, but Tonstijn and Schunovale threatened me that I would never be able to get back into Holland. In December of 1989, I left for Moscow, hoping to return to Holland soon, but the Dutch embassy refused me a visa. Apparently my quasi-partners had carried out their threat.

Nonetheless, I had to file an invention application at The Hague Patent Bureau by the end of May 1990. I explained to the consul at the embassy that I needed to claim my invention. I also mentioned de Wisser's approval and the telephone number of the bureau clerk in charge of my case. This worked, and in a couple of days I was finally granted a visa.

On May 25, 1990, I was back in Holland and immediately went to The Hague to file the papers. Yet in Amsterdam, at the Department of Labor, where I had filed my papers six months earlier, I found Tonstijn's fingerprints. It turned out that he had asked the police not to let me back into the country. Since the Department of Labor asks the police for client references, of course they wouldn't say anything good about me. What was I supposed to do? I was devastated.

Then suddenly there was a ray of hope. Dr. Maylor, whom I called to tell of my arrival, offered to resume our collaboration. I couldn't believe my ears as the neurology professor offered me a job at the university. He got a green light to set up a group to study acupuncture and reserved a place for me, too. In a few days I would report to work.

"You will have five thousand guldens for your research: four thousand for you and a thousand for the assistant or student who will be helping you, if you accept this temporary position," Maylor told me.

Longevity and Health

At the institute, I met a psychophysiology professor and spoke at the department meeting about my research. We then plunged into our work for the next few months. My assistants were a young femal doctor, a recent graduate, and a senior invited by Maylor. Before starting the project, I followed the patent bureau's advice and asked them to sign a contract stating that they could not use my method or give any information to a third party.

Upon the completion of my EAG project, I asked Maylor to write me a results report. He agreed with reluctance, and the way he subsequently played down the method's effectiveness led me to believe he was playing a game I did not understand. I asked to see a professor of neurology whose name I cannot recall.

"Why?" Maylor asked, surprised.

"What do you mean, why? I believe he has an opening for me."

Maylor raised his eyebrow in surprise. "You are mistaken. There is no opening. They are planning to create a group about two years from now."

"And what about the four thousand I was promised?" I asked, sensing something was wrong.

It turned out that there was no money.

"If this does not suit you," Maylor declared insolently, "then find yourself another partner!"

His behavior had a simple explanation. He decided he had already learned everything he needed to know from me.

There is an old Western parable about a potter and his apprentice. One fine day the latter decided that he had learned all the secrets of the trade and his pottery was equal to his teacher's. In vain did the potter ask the young man to stay for another year to finish his training; the latter wouldn't listen. Instead, he opened up his own shop. His bowls, plates, and cups sold quickly. At first glance, their shapes, bright colors, and beautiful design differed little from the teacher's. Yet the young potter saw a difference, for when he took a plate out of the oven, the glazing was corrupted by a fine web of cracks, noticeable only to the master and the most discerning buyer.

The young man changed the clay, the paint, the temperature, but nothing worked. He went to the teacher. "Teacher, why don't your plates have these stupid cracks, and I can't get rid of them?"

The Dutch Maze

"I will tell you the secret," the teacher said, "but you'll have to remain my apprentice to the end of my life."

Anyone else would have laughed and walk away. However, the young man wished to become a real master, so he pondered for a while and finally agreed. Years passed, and the master's pottery still came out exquisitely, while the student's was always covered with the fine cracks and crevices. Only before his death did the master finally reveal his secret: "Before placing the dish into the oven, blow the dust off it."

I remembered this parable not to ascribe to myself some knowledge unknown to other scholars. But it so happened that, in research of the autonomous nervous system and putting the results in practice, I could achieve what others had not. Dr. Maylor tried to continue this research on the autonomous nervous system with his coworkers. A number of times after our partnership fell apart, he and his coworkers tried to use my tool, but it never worked.

In early 1991, I found out by chance that Holland has an innovation center in Amsterdam. I was introduced to Paussen, director of medical research, who then introduced me to a businessman named Shram. This was the kind of person I needed: smart, enterprising, and energetic. He knew no more about medicine than an average person. My research interested him only in its commercial aspect. After some time, he offered to discuss the possibility of signing a contract.

In August 1991, I left for America for a month, met a lot of people there, and made a number of useful contacts. One of my new associates, Dr. Felder, tried to get me to move to San Francisco, but I wanted to stay in Holland, because in the meantime I got permission to stay in Holland permanently, helped by the fact that my daughter was a Dutch citizen. When I returned from the States, Shram had already prepared a contract and even had it translated into Russian, so that I knew what it said. The contract provided for setting up a company; I would be the technical director and Shram the financial director, who would invest money, hire people, and rent the office.

Yet when it came to signing the actual contract, everything was different. There were now two contracts, one for creating the company and the other for hiring me as his servant. I did not understand why two documents were necessary, so I consulted a lawyer. After speaking with the lawyer (mostly in Dutch), Shram said that in two days he would let me know whether he was still interested in a partnership. His words made me wary, because just two days earlier he had been seriously interested.

I was confused. What had happened? What were the motives for Shram's odd behavior? Until then, he had acted impeccably. For example, once, when I had urgently needed to extend my stay in the country, Shram had rushed to the police station at nine in the morning, never mind that his mother had passed away the night before. He had dropped everything to help me! I felt that I was losing not only a good manager, but a good friend as well. I never found out the reasons for Shram's sudden change of heart, but I think that someone had interfered in a negative way. Who? The following events shed some light on this mystery.

Paussen wasn't discouraged and calmed me down.

"Maybe this is for the better. I will introduce you to one of Holland's best managers!" This manager was Khijs Kaismonbergen, a tall, massive man. It turned out that he already knew of my work and agreed to help me. When we met, Kaismonbergen informed me right away that he had already found an investor, and it was time to make up a contract.

"Now I am your manager," he said, "and I will handle all negotiations. You are responsible for your research, and I for everything else. You are a director, and I am a director, too."

My experience told me that I should be careful, and I didn't like this idea. Kaismonbergen convinced me that this was the proper way to do business and that the director's title would make him look more respectable and make for easier contacts with future investors. He wanted to get 40 percent of all future profits. Paussen told me that this was not a high figure, considering the amount of work involved. I also had to take into consideration that I did not have Dutch citizenship, plus they would do everything to get me a job, solve all of my problems, create a name for me, and make me famous.

They were extremely persistent, promising to provide me with a job in the next four months and insisting that if I did not have a job or financing by the end of four months, the contract would no longer be in effect. Perhaps this was why Shram had rejected our partnership, I thought to myself. Maybe it was Paussen's idea to replace him with

The Dutch Maze

Kaismonbergen. It also made sense, considering that Kaismonbergen offered to make Paussen our arbitrator, should such events ever occur. He insisted that Paussen was not an interested party, however ridiculous his disinterest sounded.

The long countdown began on September 14, 1991. A month passed, then two months, three. My so-called "co-director" continued to feed me promises and confirmations. "Four months have passed, and I have no job, no residence permit, and no investor," I told him. "I think I have every right to break the contract."

After a couple of days, Kaismonbergen called me and invited me to a McDonald's to talk business. On our way to McDonald's, Kaismonbergen pelted me with questions. *Would I work with him if everything went well? Would I pay him approximately 40 percent of the profit? If I found a job in America, would I still pay him the 40 percent? Was I thinking about returning to the Soviet Union? Why didn't I want to work in Moscow?*

Suddenly, he took out of his pocket a small Dictaphone and announced that he was a secret agent and that our conversation had been recorded and now I was in his hands.

"So I suggest that you do not terminate our contract or else you will be sent back to Moscow, where the KGB will make sure you will end up in Siberia!"

"This is blackmail!" I managed to murmur.

My wife could not believe the story. Kaismonbergen had been to our house, and we had treated him hospitably. He even admitted that he felt at home with us, as if we were family.

For a couple of days, I did not want to see anyone, speak with anyone, or think of anything. I was too upset and shocked by what had happened. Scared, too, to be honest. Where could I find an honest partner? Where would I find investors?

I visited the international trade center and told one of the workers who I was and what I did for a living. He introduced me to a manager named Belsmyt, who became so enthused about my ideas that his wife said "he was pregnant with Lerner." He came to see me practically every day, sometimes even twice a day, and during one of our conversations, I shared my worries with him. My visa was about to expire, and I had neither a job nor a residence permit. Belsmyt reflected.

"Let's do this," he finally said. "I will give you a job and will pay you five thousand guldens a month. In effect, I will buy you time, and you will get a residence permit. This is at least some kind of a way out. Meanwhile, I'll look for investors."

I accepted his offer gratefully and felt it was necessary to explain to Belsmyt the entire situation, showing him Kaismonbergen's contract and explaining my wish to revoke it.

Belsmyt read this document and broke out in laughter, "How could you sign this? I'll write us a new one!"

Soon, a new draft was written, but I was in no rush to sign. I decided to consult with another businessman named Terghuys, who knew of my research and agreed to invest. He owned a company and was friends with Kaismonbergen.

When I came to his office, his wife, hearing I was from Russia, rushed to the store, bought some fish, and cooked it for me. She thought, if he is from Russia, he must be hungry. But you should have seen how upset Terghuys was that she had spent money on fish.

"I am sorry to disappoint you," Terghuys said. "If you sign such a contract, you will fall into a trap from which you will never get out. I can help you, though. We can sign a contract for four years, and I'll pay you the same five thousand guldens a month. The only difference is that you will get a regular contract. In addition, Belsmyt has nothing to do with medicine, while my company manufactures medical equipment. I'll give you two engineers, and we will start working on your device right away. Come to my company, and I will show you everything. You can then make your decision."

I liked his offer. The only unpleasant part of it was having to resume contact with Kaismonbergen, but I had no choice.

The following day, we went to his office. It made a most favorable impression. It looked modern and attractive. His company employed about ten people. Now his offer appeared even more tempting. However, as we discussed the contract, I couldn't help thinking of a Russian fable, where a man was dividing a turnip with a bear: "I get the top and you get the bottom." If I agreed to this document, I would have absolutely no rights and could even lose all my patents. Terghuys, however, had the right to change any part of the contract and even revoke it. In other words, he was the boss of the situation. Yet this was better than what Belsmyt had offered; at least Terghuys promised to begin working on my instrument immediately.

Although the conditions did not suit me, Terghuys seemed businesslike and somewhat charming. Kaismonbergen also took a part in drafting the contract and signed a separate agreement that stipulated forthwith, he would receive only twenty-five percent of the profit rather than forty.

Before signing, I warned Terghuys and Kaismonbergen that the patents would cost them a pretty penny—several tens of thousands of guldens, which I didn't have. They instantly assured me they would find the money.

"Do we need to put this in the contract?" I asked.

"No need. If a Dutchman promises something, he never lies."

In March of 1992, I started work. Terghuys introduced me to the engineer who was to help me. However, two days later, the engineer ended up in the hospital. When I asked Terghuys to find me another one, he didn't. The engineer spent two weeks in hospital. I spent the two weeks I spent. By then, a bill came for the patents, to the tune of almost twenty thousand guldens. Terghuys calmed me down. I should forward him the bill, and he would pay. I did that. I called the agency, and they faxed him the statement.

Finally, the engineer recovered and was able to work, but suddenly it turned out that there was a more urgent job for him to do. I was left no choice but to wait. A week passed, and then another week. I kept asking Terghuys, "When we can finally start?"

"Soon, soon. There is a technical college in the area, and I'll invite a few students for their field work here."

A month passed without a job or a paycheck. The second month was coming to an end, and still no progress. Meanwhile Terghuys was feeding me with promises and pathetic excuses. He said that his wife was ready to go to the bank and wire the money to my account but then remembered that neither of them knew the number of my bank account. I gave it to them.

After a few days I asked, "Did you transfer the money?"

"No, my wife mislaid your number and cannot find it. Give it to me once more."

I gave him the number once again. After a couple of weeks, another lame excuse came up. Finally, after almost four months,

I heard, "We have transferred the money to your account. You can go and get it."

At the bank, no money. What was I to live on? How long must I wait?

Finally, Terghuys solemnly stated, "Congratulations, your first salary has been transferred to your account. I hope you have received it."

I rushed to the bank—my account was still empty. After a few days, still nothing.

Barely containing myself, I said, "Mr. Terghuys, how am I to understand this? You said that the money had been transferred, but I still have not received a single gulden."

"You haven't and you won't. I sent your salary to the patent bureau, and for another four months I will be using it to pay off your patents."

"But you promised to find investors or find other sources for this! Did we agree that the payment was going to be coming out of my own paycheck? Did you receive my permission for taking care of a situation in this manner?"

"Yes, we agreed that I would forward your paycheck to the agency in order to pay for the patent," he said, looking me straight in the eye.

"This is the first time I am hearing this!" I screamed.

"What do you mean? Kaismonbergen and I agreed on everything with you, and he as your agent can confirm your words."

"How can he confirm something that did not happen? I did not make any such promise!" I protested.

"A verbal agreement is equivalent to a written one. We, the Dutch, are honest people and keep our word."

This insolence almost made me speechless.

"Don't you remember that you and I and Paussen discussed searching for investors to pay for patents?"

"No, I don't. The payment for patents comes out of your pocket."

"In that case," I said, "I no longer wish to have anything to do with you or Kaismonbergen. I have wasted several months. I didn't get help, or a salary, and I was lied to as well. That's it! I am breaking the contract!"

I did not know what to do next, so I went back to Belsmyt, who had an idea. On the day after his meeting with Terghuys, I discovered that he had decided to invest a large sum of money into our project. I had not expected this turnaround at all, but I stuck to my guns: no work until I got everything I was owed.

Terghuys and Belsmyt went on to explain that they couldn't pay me officially because it would entail high taxes. Therefore, they would "loan" me the money that they actually owed me, and I would have to pay it back only if I made a large profit from the manufactured equipment. This made sense, and I agreed. Finally the long-awaited money made it into my account.

After this, we made a new contract. I decided to revoke the old one and asked my lawyer to write a letter to Kaismonbergen to that effect. This needed to be done fast, since the contract stated that if I did not revoke it within a year, it would remain in effect for the rest of my life. However, for some reason my lawyer kept delaying sending the letter.

Only two weeks remained until the deadline, and the lawyer was still dragging his feet. There was clearly something to it. Later, he admitted that Kaismonbergen kept calling him and trying to win him to his side. Finally, I demanded an end be put to the matter, but with no results. I then let Kaismonbergen know in person that I was revoking the contract and considered our cooperation finished.

I found out that Kaismonbergen, along with his wife and his brother, had invested fifteen thousand guldens into my project. Belsmyt and his brother invested an even larger sum, and a good chunk of money came in from the economy ministry. Of course, the reason was that my work appeared profitable, and the partners could not divide the future profits amongst themselves. The only winner in this game turned out to be Terghuys. He simply put all the money in his pocket, and without spending a gulden, saved his own company from bankruptcy.

At that time, I did not know anything about this. One lovely day, Terghuys invited Belsmyt and me to a restaurant. Because of his stinginess, such an invitation was unheard of. He never spent a penny—he didn't have to—but this time, he ordered good food, wine, and liqueur. The reason for this splurge soon became obvious.

"Today, I am celebrating!" he announced. "Today my company is saved!" He said he was thankful because I had played an important role in this. What role it was he did not specify, but I understood that the money invested for my project saved his entire company.

Longevity and Health

The entire night, he was in a good mood, and by the end of dinner, he told me he wanted me to acquire aristocratic manners and learn proper Western behavior. He had a friend, a duchess, who taught a special two-day course on manners, including table and business manners, to all his workers. Her lectures were accompanied by practice sessions, so that in four to five hours, she could make a plebeian into a noble. He promised to specifically invite the duchess to meet me. In conclusion, he lifted his glass and toasted my success. On the following day he forgot all his promises. This was typical of him, always promising a lot and then doing nothing about it.

Back at work, my student and I began creating the device. When it was finally made operational, it was obvious that it was good only for demonstration and not for serious work. The electroautonomogram needed to be recorded on both hands and feet, on four channels, but our tool had only two and lacked software. With analysis, by mere observation, much data was lost. Besides, it was very unreliable and kept breaking down.

I had brief business relations with Professor Klopper, from the Academic Medical Center (AMC) in Amsterdam. He and I planned a series of highly complex and expensive trials in which several doctors and assistants participated. Unfortunately, our instrument broke down at the most critical moment. It took a week to fix it, and then, another breakdown, another repair. On another occasion, it broke down in an even more critical situation, during a demonstration arranged for Hewlett-Packard's representatives, on which a great deal depended. The reason for the breakdowns was that Terghuys tried to cut corners.

For a few years, I didn't attend a single international conference or scholarly convention. When once I requested permission for a trip to England in order to consult with a major scientist of the autonomic nervous system, the company declined to pay my expenses. I flew twice to the United States to talk to American specialists, paying out of my own pocket. I spent a hundred guldens to fly to Nijmegen, a village in the south of Holland, where I had been invited to speak.

But forget about travel, the company refused even to pay for a needed book on autonomous pathology, and I had to take it out from the library for six months. The only thing my partners paid for was the patents, because otherwise their investment would go belly-up.

The Dutch Maze

Belsmyt assumed the responsibility of my salary. Still, every month I had to fight for my money. When potential investors showed up, Terghuys kept me out of the negotiations. He conducted the negotiations, literally behind my back, asking me to step outside into the hall. If I needed to talk to him, his secretary would say that he had no time and would not let me into his office. Of course, after these incidents, I would often go home with my heart beating violently.

At times, things turned absurd. When I was working with Klopper, and we managed to achieve great results, Terghuys suddenly announced that professor Klopper had to discuss all matters with him and not with me. But Terghuys knew nothing about the autonomous nervous system or about medicine in general.

Professor Kain, the head of the gynecology department at the university clinic, once asked me indignantly, "Why did Mr. Terghuys call to tell me nasty things about your work? We are not friends, and what kind of manners are these?" This was these gentlemen's modus operandi: they would stop at nothing to soil my image in front of my colleagues and show themselves to be leaders.

Once, in a moment of honesty, Belsmyt told me that the boss had a plan. Under some pretext, he would revoke the contract guaranteeing me a job, and after my forced departure he would acquire my patents, licenses, and, most importantly, the rights to manufacture my machine.

This was the kind of person I was forced to work with. Tied up in their contracts, I felt like a fly caught in a web, beating my wings desperately and yet unable to free myself.

Still, the work went on. At Holland University, I met a professor, whose name I don't remember. He presented my electroautonomograph to the general medicine department and invited me to speak at the presentation. But the head of the department gave me only a few seconds to speak about the tool. His whole demeanor, his body language and tone of voice seemed dismissive, as if to say, "Some Russian come to teach us!" I felt very insulted, but I could not refuse the offer.

On the scheduled day, about twenty people came—not just general specialists, but neurologists, physiologists, cardiologists, and other specialists. The presentation lasted from six in the evening till midnight, as I spoke and fielded nearly fifty questions. Although I answered every question, I was still being told, "No, we cannot agree about this." Finally, I could not hold back.

"Tell me, please, where did I make a mistake, and what part of this do you actually disagree with? I would be very grateful!"

The director of the seminar said, "We just don't think this way in Holland." I realized that further discussion was useless. The contact between me and the groups of doctors ended there and then.

While my partners did not have enough money to fully finance my project, they would neither let it go nor allow it to grow and yield results. At first, serious investors showed interest in working together, but then they felt the disagreement among the partners at Terghuys' company, lost confidence, and pulled out. This situation repeated itself. Companies like American, Texas Instruments, and Dutch Delft Instruments considered buying exclusive rights and then changed their minds.

I asked a Dutch attorney to look at my contracts. He told me there would be no talk of partnerships and collaboration until I revoked the contracts. I decided to follow his advice, and in January 1994, turned to one of the biggest law firms in Holland. However, nothing happened for five months. Although I paid a sizable advance, nothing progressed beyond conversations and letters. I remembered Terghuys' words: "I would not recommend that you take us to court. I can always prove that you put my company on the edge of bankruptcy by stressing out your coworkers and interfering with business. Kaismonbergen will definitely take you to the cleaners. Keep in mind that any court will blame you for everything."

The situation was made more difficult by the fact that I ultimately needed to get the residence permit. There was no reason for my application to be rejected, but a year passed and the Justice Ministry would not grant me citizenship, allegedly because some papers had been lost. I suspected that Terghuys and Kaismonbergen had had a hand in it, since my deportation was in their interest. Two years had convinced me that these two gentlemen were capable of anything. I did not see any way out.

Chapter 9

The Siege

I went abroad to continue research, to Belgium, Germany, and the United States. Neurologists working in the hospitals I visited regularly attended international conferences. While I continued my research, my wife, back at home in Holland, became ill with an illness that the hospital neurologist found very difficult to diagnose. I was able to make the first real diagnosis myself, and this helped my wife recover. However, some of the Dutch neurologists, including the professor, did not like me because they had made mistakes, and I showed them how to cure her. My relationship with some Dutch doctors deteriorated, although I also knew many good doctors in this hospital. In a similar way, I noticed how my relationship with colleagues from other countries went sour; I experienced what it felt like to be a persona non grata.

For example, at a conference in London, I met Professor I.G. Mayhew,. Professor Mayhew was from Edinburgh University, where Dr. Ian Wilmut had recently cloned Dolly the sheep. Professor Mayhew is a famous veterinarian who specializes in horses. He asked me if my electroautonomograph could record the skin potential of horses, as no one had ever been able to do that.

According to the professors Mayhew and others, such results would help reveal the source of the horrible "grass illness" that thousands of horses die from every year in Southern England and Scotland. This disease usually shows up in the spring, in May, and results in heavy diarrhea, along with pain so tormenting that the poor animal is euthanized out of mercy. The cause of this disease, a toxin, ends up in the horse's organ system through grass fodder and corrupts its autonomous nervous system.

After developing a test schedule and making up special electrodes, we succeeded in recording a horse's skin potential.

Professor Mayhew was thrilled. He wrote that his recent success would be the first in veterinary history and expressed hope that the Scholar Council of Edinburgh University would approve of continuing these experiments (this letter is also included in the appendix).

There is a famous professor in England named Mathias, a highly influential man with many connections. He was the president of the European Association for Autonomic Disorders. I got in touch with him and went to England to demonstrate my electroautonomograph. After being shown the device, he was very excited and showered me with compliments, saying that this was just what doctors had been dreaming of for years. He wrote me letters of recommendation and provided a few names of specialists on the autonomous system in Europe and America, with whom I was to get in touch. He also suggested that I come back in a couple of weeks and promised to give me a group of patients on which we could test the tool. With the results obtained, I would be able to write a chapter in his frequently republished manual on the autonomous nervous system. He was interested in my work, and I went home very pleased.

After two weeks, like a bolt of thunder in a clear blue sky, a letter came from Mathias with a request not to use his recommendation. He said that the situation had changed, and he had problems with my device.

"This needs to be postponed," he said. "We will get together another time and discuss this issue." I did not know what to think. Professor G. Mayhew, too, abruptly cut off his contact with me.

In 1995, at the Department of Large Animal Medicine and Nutrition (Utrecht, The Netherlands), R. A. van Nieuwstraat and others performed positive experiments on horses that showed a potential new way to diagnose grasses sickness in horses, and the doctor gave me a positive written report (also included in the appendix).

A few years passed. I kept running into Mathias at international conferences twice a year, in the United States in October, and in Europe in April. Each time, he told me, "Yes, yes, I remember you, and I am very interested in your work. We will talk later. I will set aside the time to meet with you." Then he would disappear again. When I offered to come see him in London, he would always be too "busy" to meet, but explained that he might be free in about six months. Every time I tried to find out what the problem was, he kept dodging the issue. Mathias's change of heart toward me became known to specialists from Italy, Belgium, and other countries.

Whenever I come to a new country, I meet with professors and demonstrate my data, and they are interested, yet I still don't have a university or a hospital supporting me. The specialists probably check my references. They call up people in Holland and ask, "Who is Dr. Lerner, and where did he get his data from?" I believe that the answers they receive from Holland are of a completely negative nature. Some of the Dutch doctors never forgave me for interfering with the treatment of my wife. She was ill with Guillain-Barré syndrome (a form of paralyzing disease), and recovered with my help. How else can I account for the fact that, time after time, the contacts that begin so well inevitably come to nothing?

The EAG was presented and tested with very good results, both in the Belgian Gent University by Professor Vondenstraten, and in the Italian Bologna University by Professor Pietro Corteli, along with other European hospitals. After demonstrating it in Tennessee in 1999, I also got a good report from Dr. H. Rashed, who was the director of the Autonomic Function Unit at the university (documentation included in appendix).

The same happens with investment companies. At first, everything runs smoothly, but at some particular moment, someone wants to know the opinion of Dutch professors and doctors. When potential investors receive the information, they immediately cease contact. This problem followed me for years and still today I sometimes experience the same problem, but I can't do anything about it.

Something similar happened in Belgium. As I was setting up research at the University of Gent to deliver medication in the brain via the nasal cavity, one of the doctors asked me why I could not do this research in Holland. She told me that she would visit Amsterdam soon. After she returned,she started treating my work negatively, and she refused to carry out the planned experiments. All the contracts were suspended. Apparently, she had received some information from Holland that influenced her thoughts and decisions.

In many countries, I was blocked by a group of Dutch doctors who discredited me by spreading negative information behind my back. I could not say anything in return. No one asked me what had happened or why the negative opinions existed; potential business partners would simply cut off contact and rip up our agreements.

In 2002, a professor Hilz from Germany approached me at a conference in the United States. He was well known in his community, and in other countries as well. He is active in research, makes presentations at conferences, and works with many doctors in America and many other countries.

"Dr. Lerner," he said, "let's have a beer. I see that at your age"— I was seventy-three at the time—"you are still active and attend all the meetings and conferences. This is most laudable."

Once we settled at the café, he asked me what I was currently researching.

Although I knew that telling too much is dangerous, I still told him a few things.

"We could work together." He proposed a topic for research.

"I would be pleased to be your partner, but I have already made an agreement with a professor from one of the biggest universities in America," I told him.

Hilz didn't say anything, but I felt anger coming from him. My intuition was right. The following day, when he saw me, he assaulted me with screams, accusing me of not having enough data, and on and on. Professor Hilz was apparently angry because we could not cooperate.

Just recently my colleague Oleg from New York, whom I knew from Chernovtsy, introduced me to his doctor friend in the belief he might be interested in my work. One of his first questions was about professor Hilz.

"You know," he said, "he is the king in science." I said I knew the name, but our relationship didn't work out. That answer decided my fate, and our meeting came to nothing. Moreover, even the friend who had set up the meeting came to treat me coldly and no longer wished to work with me.

I was stuck in the crossfire. On one side, I was discredited by the people who wanted to steal my ideas, and on the other hand, I was haunted by the negative attitude of some Dutch doctors I had been forced to complain about.

It had turned out exactly like that. The first conflict was with the neurologist Maylor, whom I mentioned earlier. He wanted to steal my device for research of the autonomic nervous system, but I did not let him.

The Siege

Another reason for my growing isolation was that I had a job working for Terghuys that sponsored my research in Holland and gave me a salary. Thus, my wife and I could go away on weekends, stay at good hotels, and dine at fine restaurants. I might enjoy socializing with a few immigrants, but not one of them had a job. All lived on welfare and saved every penny. When they traveled, they carried a thermos with coffee and sandwiches and stayed at the cheapest hotels. Hence, it was hard spending time with them. They exposed their envy toward me. In truth, I don't have anyone with whom I could socialize.

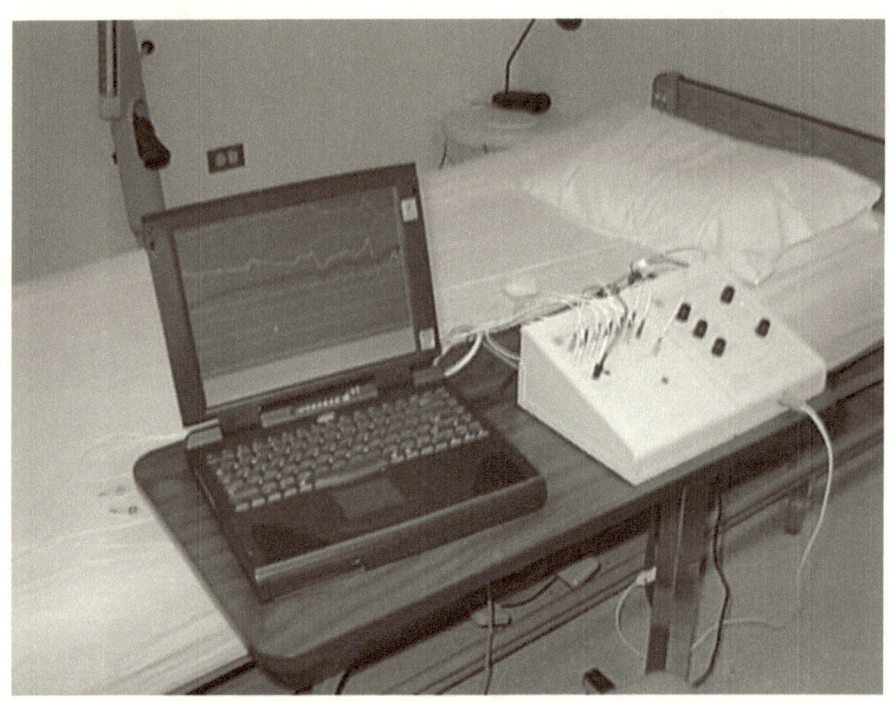

My first invention in the Netherlands was the EAG, in 1991. In the period before 1991 the EAG apparatus was much simpler and not market ready.

The Siege

The autonomic nervous system of a horse being tested with the EAG in 1999 by myself in the presence of Professor I. G. "Joe" Mayhew from the Department of Veterinary Clinical Studies at the University of Edinburgh, Scotland.

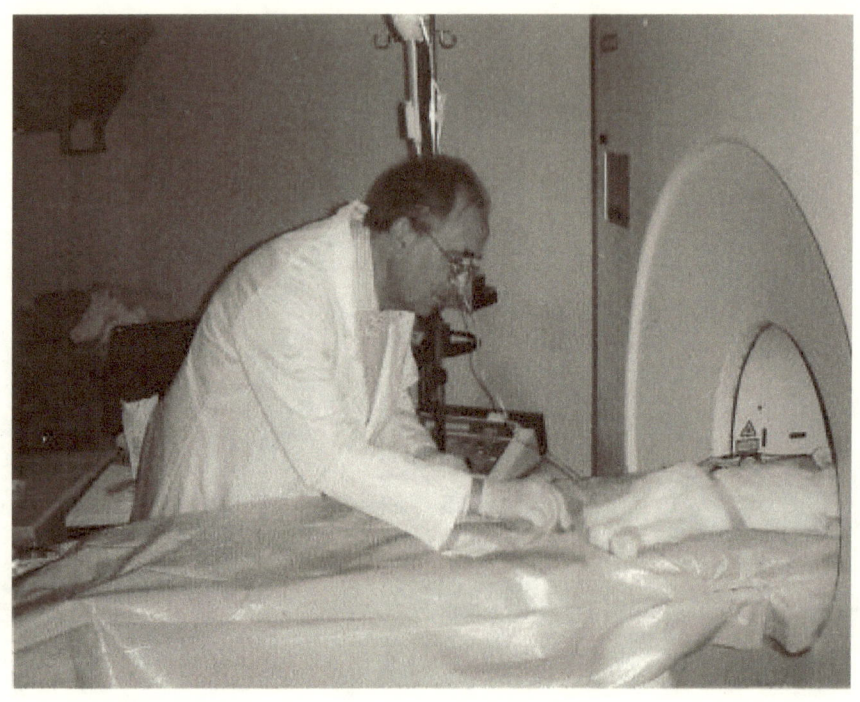

*Myself experimenting with intranasal drug delivery on a rabbit at the
University Hospital in Leuven, Belgium, in 2003.*

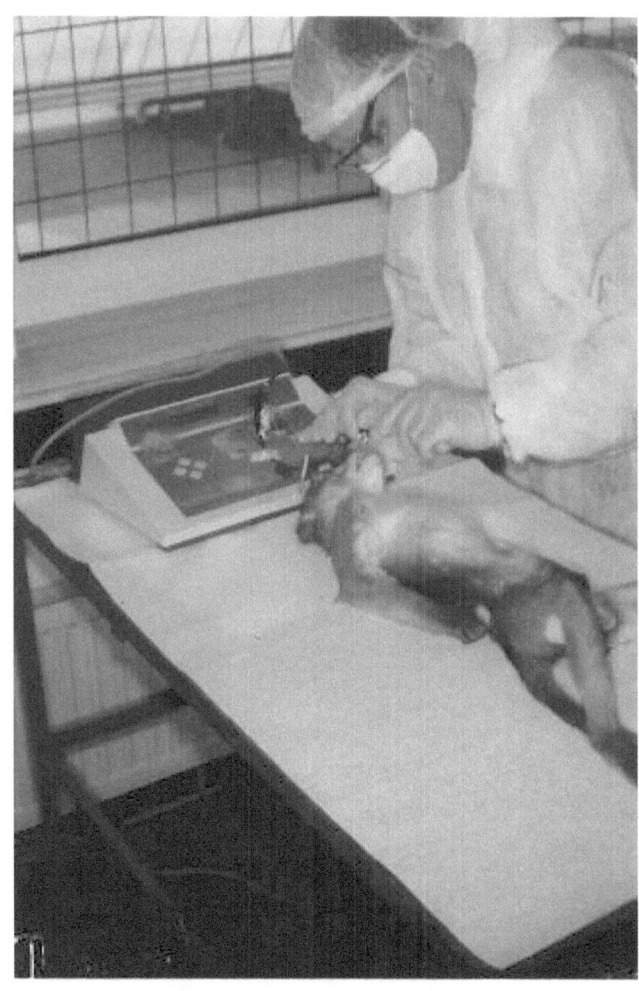

Myself experimenting with intranasal drug delivery on a monkey at the Covance Laboratories, Munster, Germany.

Dr. E. Lerner (left) receives the prize of the city of Geneve
from the mayor Alain Vaissade

Here I receive from the mayor of Geneve the prize for EAG - the Silver Plate

The Siege

*This items are the Silver Plate for EAG, one Gold Medal
for EAG, a bronze medal for ETDDS and a gold pen
from the mayor of Geneve.*

Giving a presentation in 1998 in the United States of America.

Chapter 10

Exit from the Maze

I presented my autonomograph at the world invention exhibit in Geneva in May 1996, along with my project on delivering medicine into the brain via the nasal cavity, using iontophoresis. As I have stated before, my personal invention for EAG was awarded the Gold Medal, the highest award for inventions at the Geneva exhibition. My delivery method won the Bronze, and in addition to the international awards, I also received a Silver Plate from the City of Geneva and a gold pen from the mayor of Geneva. The mayor wanted to throw a dinner in my honor, but other members thought that this was a bit too much, because I'm only a scientist from Russia.

At that time, at the World Invention Exhibit in Geneva, I met Dr. Benkendorf, a German scholar who studied AIDS in Africa. Later, in Frankfurt, he introduced me to Dr. Klaus Graf. We seemed to make a good impression on each other and felt a common bond. I immediately warned Dr. Graf that my hands and feet were tied by the TLB Electronic contract, and that the company was ready to sell the project for half a million guldens.

Dr. Graf said he would think about it. However, when his lawyer, Shoolman, started to discuss the details with Terghuys, the latter suddenly claimed he had invested half a million in the project himself. In response, Shoolman offered another hundred thousand, but Terghuys greedily demanded seven hundred thousand. The negotiations hit a dead end. In the period after we met for the first time, Dr. Graf and I stayed in good contact, and from time to time I asked him for advice.

In thousands of years, people have developed only a few ways of introducing treatment into the human system: through the skin, intravenously, in the muscle, and in the stomach. It would be helpful to contrast the different traditional ways of delivering medicine to the brain (i.e., getting around the blood-brain barrier). Three additional techniques include neurosurgical injection into the parenchyma, which has only limited diffusion into the central nervous system (CNS) tissues; pharmacological techniques that improve lipidization or lipophilicity, which may be of only limited use for large molecular weight proteins; and molecular Trojan horse techniques, which, unfortunately, involve the chronic stimulation of immune response against the introduced agents. I invented the alternative of intranasal delivery. It is hard to invent something new in this area, but by finding a way around the blood-brain barrier, I succeeded. Through my tests on animals I delivered the following medications into the brain: methylprednisolone to treat multiple sclerosis, encephalitis, meningitis, and brain hemorrhage; methotrexate to treat brain tumors; L-dopa to treat Parkinson's; tacrine to treat Alzheimer's, and octreotide, from Novartis, for a whole array of conditions.

I put the medicine directly into the nasal cavity and delivered it to the brain with a low electric volt—which is basis for iontophoresis. Iontophoresis has been used for many years to deliver drugs into the skin, the stomach, and other internal organs, but nobody had delivered medicine through the nasal cavity to the brain in this manner.

I tested these five different medicines on rabbits and monkeys and obtained excellent results that can be seen in my 2002 study. According to Professor Booij a famous from the University in Nijmegen in Holland, my research would cause a revolution in medicine. Thanks to his enthusiastic review, NOVU, the Dutch inventor's association, gave my project the premiere national award, called the "DUTCH TOP INVVENTION". The award was presented in The Hague, with hundreds of people in attendance. The description of the award read that the method of delivering medicine into the brain through the nasal cavity was no less significant to medicine than the invention of antibiotics. Although the method of delivering drugs through the brain had been known for many years, that method was used without iontophoresis. The combination of the two methods was what made this invention very interesting. In chapter 11, I will describe the drug delivery with and without iontophoresis,

Exit from the Maze

demonstrating that the drug delivery with my invention is many times more effective. In the former Soviet Union, some articles were published about drug delivery through the nasal cavity using iontophoresis. But currently there are no publications about experiments showing the proof that the drug reaches the brain.

Being presented an award with the title "Dutch Top Innovation 1997"
by the board of NOVU, the Dutch association of inventors. In the middle
is Mr. W. Pijzel, managing director of NOVU.

Later, NOVU's director called me and said that Dutch TV wanted to do a story on me. We agreed on the specifics, and I asked him if my "doctor friends" might interfere with the story. He laughed and said that I still had a Soviet mentality; such a thing couldn't happen in Holland. Nonetheless, the night before shooting, the reporter called me and said that the story had been cancelled, since Dutch doctors "are not familiar with my work" and would not be interviewed for the story. Without them, the director said, the story was pointless.

At first, Professor Booij did not believe that medication could be injected into the brain through the nasal cavity. However, since he knew me, he decided to do a few tests. We made a plan together, and soon, I was doing tests on rabbits, injecting methyl prednisolone into the nasal cavity and applying iontophoresis with a special tool. After a week, the doctors were astounded by the results: the concentration of the medication in the brain when using my method was ten to fifty times higher than if the medicine has been introduced intravenously.

When the result was received, the chemist who actually measured the concentration of the medicine in the brain sent me a fax late at night that said, "Congratulations on amazing results. Professor Booij is stunned. This experiment is worth millions of dollars. Be calm, and beware of pharmaceutical companies. We will introduce you to a serious and experienced company who will be interested in this experiment. This is a second Russian revolution!"

After such a glowing result, Professor Booij, promised to make up a patent application in my name. A couple of days later, his assistant handed me a few pages and suggested I should submit the document ASAP in order to "keep the patent date." At that time, I didn't know anything about patenting, my English was poor, and I asked Booij whether I had to review the application. Professor Booij told me that he and his colleagues had tremendous experience in the area and I should trust him.

A few years later, I received a patent for nasal introduction of medication by iontophoresis. When I saw the description of the procedure, I felt sick. Not only did the description have nothing to do with my method, but it abounded in errors, incorrect calculations, repetitions, and illustrations that had nothing to do with the subject. This parody of a patent discredited both my method and me personally. It took me a long time, with legal help, to change this deliberately poorly-made patent. I published an appendix that corrected the errors in the description. Only later did I realize that its purpose had been to steal my method.

Professor Booij suggested that I sign a contract with his university. The university's attorney drafted a partnership agreement for a three-month period. It stipulated that a large sum of money would be earmarked for testing, and I would not have to pay it back. In the meantime, the lawyer, who was also the coordinator for grants and contracts, didn't understand that a patent belonged to the author who had done good experiments and not just someone who had published a paper.

Three months passed, and we did not perform a single test. All this time, I kept trying to get Professor Booij and the university to live up to the agreement, but nothing happened. Finally, I decided to free myself of the university and Professor Booij as well. My relationship with Nijmegen University ended, and to this day I have no idea what they are doing with my method.

I was introduced to a lawyer, a relatively young man who charged very little, only seventy-five guldens per hour, provided that the client contracted to use his services for two years. I was tempted by the low rate. When the lawyer went over my claims regarding the university, he assured me that this was a done deal. I was in the right, and the law was on my side.

However, once he consulted the university, he started acting completely differently. Now he tried to assure me that this was a very respectable university with serious personnel. He gave me a blank agreement, not filled out and in Dutch. He took my hesitancy to sign a blank piece of paper as a display of bad faith. There we parted ways, the end of yet another contact with a lawyer.

Soon another lawyer came along, Mr. Lonstijn. He was recommended by my treating physician, Dr. van Coevorden, and I was already acquainted with him, as well as his parents. When my problems came up, Lonstijn said he had a friend who was very knowledgeable about contracts. His name was Frotauer, a person with many connections and possibilities.

Frotauer assured me that he had a true genius employed in his company, a man who had never lost a case. "Of course, he is very busy. It is hard to get a meeting with him, but it is worth a try."

We went to a village up north. Upon entering his poor, ill-kept apartment, I immediately felt dubious. The host seemed embarrassed, too, but Frotauer lauded him and his talents so effusively that he soon stopped doubting his own abilities. I explained my case, and he promised to smooth things out easily and quickly.

A week later, Frotauer called me.

"Dr. Lerner, you can dance with joy, because my friend has already discussed your case over dinner with the president of Bayer Corporation!"

For those who know who's who in the pharmaceutical world, the name of this German giant is equivalent to mentioning "Ford" to car lovers, "Hollywood" to movie fans, or "Beatles" to music lovers. Bayer's sales run to tens of billions of dollars. And here I had been given a golden key to the door of the pharmaceutical empire!

Alas, it was too soon to celebrate. At the meeting with the company representative, we reached an agreement stipulating that everybody would get a cut: my physician, his lawyer friend, and his friend's friend. Then Frotauer hinted that as far as profit was concerned, we could manage without the doctor and without Lonstijn.

"But they were the ones who recommended you!" I said. "You said that Lonstijn was your best friend!"

His answer was basically this: friendship was friendship, but business was business. This hypocrisy stung me. In addition, Lonstijn was working on two of my cases—my relationship with Terghuys and my stamp collection, which had been somewhat damaged after a flood in my house, caused by broken pipes. Luckily, by that time I had already sold the biggest part of my collection, so only a part of my collection had been damaged.

The house owner would not even hear of compensating me for my losses. I went to court. Unfortunately, Lonstijn handled the case so poorly that I lost. In fact, he did everything he could to make me lose the case. I turned to another lawyer, with the same result. I dealt with many lawyers in Holland, spent a ton of money, but never won a single case. I may allow that they were good lawyers, but perhaps in a very special sense that I could not perceive. Perhaps they didn't try very hard to win the case for a "Russian", but they got paid anyway.

After my triumph in Geneva, I wrote to a pharmaceutical company in Basel, Switzerland, called Ciba-Geigy (the future Novartis). They were interested in my method. After they studied the materials I sent them, they invited me to Basel for a meeting and negotiations. They also asked whether I would make a presentation, for which I would be paid 60,000 Swiss francs.

As far as I could tell, the report impressed the audience. However, I was impressed even more when I mentioned the fee afterwards.

"Do you have this offer in writing?" the manager of Ciba-Geigy asked.

Exit from the Maze

You have here a representative of the company, I said, who asked me to do this lecture and can confirm his promise. However, the representative kept mum. Like a naïve child, I was told that if the promise was not in writing, I would get nothing. It was hard to get used to the fact that a respectable company could conduct itself like a shell-game con man.

At that time the pharmaceutical company from Basel, Ciba-Geigy, heard about my experiments with Professor Booij and sent some expert to see Proffesor Booij who suggested that I demonstrate my method. I agreed, but learning from bitter experience, I first asked the company for a written promise that they would not use the method, for which I had exclusive rights.

The expert doctor, Dr. van Hoogevest, from Basel, wrote the promise in English. This was another lesson for me. When I received the promise from him, written in English, I naïvely assumed that the letter said the same thing I had been promised orally, and I never bothered to read it. Yet all it said was that such-and-such came to see my experiments, and the company would decide what to do next. Not a word on any promises made! Only after the demonstration did I smell a rat. It appeared that this gentleman from Basel—like Maylor the neurologist—decided that he knew everything already.

And then I suddenly discovered that professor Booij, who had still not fulfilled any of the requirements of our contract, was secretly negotiating with this pharmaceutical company and even drove to Basel with the university's lawyer. He came back and offered me a new contract to sign. This time, I read everything very carefully and was struck speechless.

What was my role in this project, according to these gentlemen? Per the contract, the company had no responsibility to me and could break the contract at any moment. At the same time, I could break the contract only at the price of giving up all my rights to my invention. Signing such a suicidal document was out of the question. For me, it was enough that neither the company nor Professor Booij had lived up to his obligations.

When I asked Professor Booij, what he personally was hoping to get out of our partnership, he said he was counting on the maximum.

"What do you mean by that? I believe your maximum is 50 percent of the profits," I said. But the professor wanted at least 95 percent.

121

Longevity and Health

When I told Klaus Graf about this, he advised me to return the money to the university and demand a paper stating they had no claims on me. I did that: I wrote Booij a letter and enclosed a check for part of the money and promised to repay the balance if he confirmed in writing that he had no claims on me. I never heard from him and never got my money back.

In Frankfurt, Dr. Graf and I signed an agreement of partnership under which I would receive 65 percent of the profit, and he would get the remaining 35 percent. Upon signing the contract, he wrote me a check for seventy-five thousand marks for development of the project. The partnership was focused on two separate projects: the development of a diagnosis device, called electroautonomograph (EAG), and the development of a new method that supported enhanced transnasal drug delivery system (ETDDS). The money allowed me to pay the patent bureau, but, most importantly, that day marked the beginning of our partnership, which has now been solid for ten years. I had finally found a businessman whom I felt I could trust.

Dr. Klaus Graf is a businessman who operates all over the world: Europe, Asia, Africa, and South America. His interests are not limited to medicine. For example, he co-owns a soccer team in Spain. He spends most of the year at his villa outside Zurich, where he also has an office. He works with a large staff, but we always had a woman named Mrs. Weiss working with us, a lawyer who is also Dr. Graf's assistant.

And so, at long last, I felt that I had found a solid base from which to build, test, verify, and market my inventions. One of the first things to do was to decide where we would create our new company and what it would be like. We discussed all the aspects of this business for a long time and eventually decided to establish Lerner Medical Technology Limited, which was to coordinate research, business contacts, management, and two more companies in the Dutch Antilles. We registered two companies there, Prognomed (prognostic medicine), which would deal with electroautonomograph, and Intrabrain, which would carry on the development and implementation of delivering medicine into the brain with iontophoresis.

The method I have discussed before of delivering medicine to the brain through the nasal cavity had been used in the USSR for years. Yet when I repeated these experiments on monkeys, I discovered that instead of the brain, as Soviet doctors thought, the medicine ended up in the bloodstream. In Holland I patented a method radically different from the one used in Russia.

Klaus Graf invested money into this project, agreeing to finance the company as needed and expecting to make the money back from future profits. I became the de facto manager, since was Graf too busy and sometimes took a couple of weeks to get back to me. It was hard enough to even meet and discuss business every two weeks or so.

My office in Amsterdam is located next to my house. It has a large room where meetings can be held, and rooms that are well-equipped for research. The first four years, I had a secretary. I had two coworkers to assist me, graduates from the pharmacy department at Leiden University. One dealt with the electroautonomograph, and the other, with delivering medicine into the brain. It looked as though I couldn't ask for more. Yet problems soon arose, and most importantly, the relations with the coworkers went sour.

In the beginning, everything was fine, though I couldn't help feeling some enmity on their part. They were often surprised that everything I did turned out so successfully. In the university, when an idea is developed, the entire laboratory spends six months in preparation, reading and analyzing the literature on the subject, going over all the details, consulting with other laboratories, until finally, the experiment is done and the result comes out negative!

We used homemade electrodes and equipment, and conducted the experiment without any lengthy preparation, yet the results were positive! When one of my coworkers began to argue with me, the other would say, "Do not argue, because in the end it happens exactly as Dr. Lerner said it would. If he said it will work, then it will, and if he said it won't, then it won't."

I gradually began to notice that, for reasons I did not understand, they were slowing down their work and deliberately refusing to follow my requests. I think the partial explanation was that, though the experiments were based on my ideas, I was completely dependent on their integrity and professionalism. They could not help being influenced by the atmosphere of enmity around me, though perhaps they understood where it came from. I completely depended on them due to my lack of knowledge of Dutch and computer technology, and they could not deny themselves the pleasure of taking advantage of my shortcomings. In time, this led to complete rudeness, failure to follow

123

my requests, and concealment of important letters and bills, which resulted in financial problems. Writing about some of their actions still causes me pain.

Dutch laws guaranteed that my subordinates would receive complete social protection. I could not tear up the contract and discontinue their services. On the other hand, who could I hire to replace them? No matter who would take their place, the situation would be the same, because the circumstances were the same. I was forced to tolerate them, even pander to them, in order to make them work. All this put me in a state that made it impossible for me to be at the office; I felt so sick that I would drop everything and leave just to save myself from a heart attack or a stroke.

Once, I accidentally read an e-mail that my secretary received from one of my employees. "Please tell the old Russian to keep his Communist hands away from the computer." The strength I could summon to cope with these people's boorishness convinced me of the importance of my research and gave me confidence in its results.

I finally got rid of them, but I could not work by myself, and I needed to look for a new assistant. Someone recommended Givi, a young man from the Caucasus, an engineer who had lived in Holland for seven years and spoke the language fluently. He worked for me for about six months. True, he was a good engineer, but soon I cursed the day I had hired him. Every day, he was an hour and a half late; he left work early and made up absurd excuses. Now he was sick, now his car broke down. As soon as I left the office, he would either disappear as well or would work on his own projects. Early on, he took all the laptops home for a few days. I don't know what he did with them, but his answers to my inquiries didn't make sense. Upon his return, the computers started malfunctioning. At first I had no suspicions, but I started having them when an older coworker told me he could not find a special electrical extension cord. I looked through every centimeter of the office space, but the extension was nowhere to be found. I asked Givi,

"Where is the extension cord? It was in the office."

"What do I have to do with this?" he screamed. "It's not my fault!"

Yet the next day the extension cord magically materialized on my desk. Now, it occurred to me that he was friendly with computer engineers from the former Soviet Union who had come to the office in order to "improve" our computers' performance. I had my suspicions, but I was unable to prove anything. The most horrible discovery was still ahead of me. For six months, Givi assisted me with experiments on monkeys. Their essence was that after the injection of medicine into the brain, the animal would be put to death, and certain parts of the brain were placed in containers for research. These experiments were very expensive—almost twenty five thousand dollars apiece—and their results affected the research.

What did I find out at the end? It is hard to believe, but Givi had intentionally placed the samples in wrong reports. Of course, I fired him, but the experiment was ruined, and six months of work and lots of money went down the drain.

Much later, I found out that two electroautonomographs were missing, and a couple of others were damaged. Givi knew the local laws and I could not prove anything. True, he once told me that his father was a doctor and was currently doing research in the Caucasus in alternative medicine. Perhaps he sent two machines to his own father the researcher? I cannot find any other explanation. But I drew a conclusion: no more hiring former Soviets, even if it solved the language problem at work. As for experiments, I had to repeat them with some help from Elske van Zanten, and I received exactly the results that I had been expecting.

But back to work. I discussed all the legal problems with Shoolman, Dr. Graf's lawyer, who lived in Boston. After ceasing my connection with Terghuys, I was forced to break up with the patent bureau in Hamburg, and I transferred my patent cases to Boston. The Hamburg bureau patented my work in Europe and Japan, while the Boston one dealt with the United States, Canada, and Mexico.

One of the most important goals was to find a reliable manufacturer of the device we had been developing all these years. Finally I chose Twenty Medical System, a small Dutch company that developed medical equipment and supplied such major firms as Phillips and others.

In order to receive more complete information about the condition of the autonomous system, I suggested that we make the device with eight channels. To the four channels that record the skin potential we added another four to record cardiograms, electrogastrograms, blood pressure parameters, and blood oxygenation. Besides, the new device had to be equipped with a stimulator producing sound and electric impulses—painless, but sufficient to evoke a reaction.

Within a year specialists from Twenty Medical System manufactured ten such devices, which worked perfectly under any conditions. At the same time, we continued the research on delivering medicine into the brain in several laboratories, in order to guarantee the results to be independent and objective. The experiments done on rabbits continued in Hamburg in one of the oldest pharmacological toxicological laboratories, led by Dr. Leishner. The concentration of medication injected into the rabbit's brain was studied in a biochemical laboratory in Dortmund.

We chose, as I have already described, several tools for experiments: methylprednisolone, tried earlier in a Dutch University and producing encouraging results; L-dopa, used for treating Parkinson's; tacrine, for treating Alzheimer's; and methotrexate, for treating brain tumors. In the last case, when methotrexate is introduced via IV, it does not get into the brain. Our experiments displayed the effectiveness of delivering the medicine into the brain, since the concentration in this case is higher and side effects are insignificant.

It was also necessary to confirm whether iontophoresis harmed the nasal mucous membrane. For this purpose, samples of nasal mucus frozen in Dr. Leishner's Hamburg laboratory were sent to a histology lab in Munster. The results officially confirmed that iontophoresis was not harmful to nasal mucus and had no harmful structural effect.

An American professor named Richard Guy consulted us on all questions related to iontophoresis. He currently works in Switzerland and runs an international laboratory on iontophoresis. Through him we obtained two very important results: the new method of delivery was both effective and harmless. The laboratories recommended that we start clinical tests on humans as well.

I was not able to continue doing experiments on the autonomous nervous system, because I did not have a license to practice medicine in Holland. For that reason we tested the electroautonomograph in Belgium, at the famed Ghent University, in Professor Vandenstraten's Center for Rehabilitation and Physical Therapy. We tested the device twice and made a very good impression. I also received a positive response, as I have described, from the famous Medical Professor Cartelli from the Bologna University, one of the oldest in Europe. Since 1088, it has been attended by such famous students as Dante, Petrarch, Kopernick, and Durer.

I started my research in Moscow and continued it in Holland; I was joined by scientists from Germany, the United States, Belgium, Italy, and Scotland. This work has been presented at more than ten international conferences in Europe and the United States. Besides that, we prepared an article for *Clinical Autonomic Research*, a journal published by the American Neurologists Association that deals with the autonomous nervous system. The association was founded in 1992 by Professor Robertson of Vanderbilt University.

Professor Robertson, to whom I was introduced in 2000, displayed a great deal of interest in our work. For research, his coworkers, Professor Diedriecht, Professor Biaggioni, and intern Laura Veskans, selected three groups of patients with different autonomous nervous system disorders, as well as a control group of healthy people. Biaggioni had a problem with me for some reason I still do not understand, to this day as I had a rather good relationship with the man. My American colleagues conducted experiments for seven years. Their conclusions confirmed my belief that the electroautonomograph should gain prominence in the research of the nervous system and that it is needed not only by neurologists but by other specialists as well. However, these colleagues excluded my name from an article that they published in a magazine. I had given Professor Diedriecht the computer program for the EAG because I knew he was very clever and because I expected him to be honest. Earlier, he had asked me to sell him the software for three thousand dollars, yet still I provided it for free. After this I heard nothing more from him.

A programmer and me continued to work on the device, enhancing its reliability and improving the design. Further, we developed a special computer program at the University of Karlsruhe, in Germany, that not only allowed it to register but also to analyze the obtained data—for example, to assess the level of damage in the brain and determine whether it is central, two-sided, one-sided, or local damage. We had come a long way since the first two-channel device that I had built at Terghuys's company.

Sometimes I could not believe that I had completed such a huge task and actually created this device!

As one American professor told me, anyone can make statements about their inventions, but people would require proof of what I was saying. He told me to publish all of my patents, presentations to international symposia, and any other relevant material that could provide proof about my inventions. I followed his advice.

Everything seemed to be going along well. I was introduced to the vice president of a giant American corporation, Johnson & Johnson, a man named Mr. Staf van Reet. He appeared to be a gentle person and promised to introduce me to a former employee, Mr. Brugmans, who had recently retired but was still very active and had significant connections. He could be a manager in the realization of my project. I almost lost sleep over the huge opportunity. Finally, we met.

I had an unfortunate incident with Johnson & Johnson as well. A Dutch doctor I knew introduced me to their consultant, Dr. Wise Young, professor and director of the Center for Collaboration in Neuroscience at Rutgers University. At our meeting, I showed him two of my projects, the electroautonomograph and the method of delivering medicine into the brain.

The professor and his assistant were not merely interested but called the ETDDS system, one of the most important inventions in medicine. They agreed to allow me to start research in their laboratory right away. This was interesting because in Holland, Mr. Graf provided me only with an office and not a laboratory. The manager from Johnson & Johnson also agreed to invest right away. I was to arrive two weeks later, and we would start working. This interest can be explained by the company's long-term attempts to deliver medicine into the brain without surgical intervention. Usually, they did the trepanation of the brain, and the medicine was injected through the ventricles—complex neurosurgical procedures.

I remained elated for the next two weeks, anticipating interesting work. In the third week I received a message saying that the research needed to be postponed for one or two months because they needed more time for preparation. In another month, another notice came, this time postponing the project for four to five months. Finally, six months later, the assistant of professor Wise Young told me that the original plan had changed, and Proffesor Wise Young had decided to hold off on my research. They promised that in the future they would definitely return to my project. I could have understood this if we had performed a battery of tests with negative results, but as things stood, I can only conclude that one of the companies that belonged to Johnson & Johnson decided it was not worthwhile to experimentally contrast the efficacy of intraventricular / intrathecal and transnasal delivery systems.

Exiting from the maze sometimes gets lighter and sometimes goes dark again. Some businessmen have still shown interest in my ideas, and my work on delivering medicine into the brain continues. Developing a new medication is a process not only long and difficult, but also requiring a tremendous amount of money. It takes from eight to fifteen years to create a medicine, and during those years the company invests hundreds of millions of dollars into the product.

The creation of medicines, the laboratory and clinical experiments the manufacturing, the marketing—this is a colossal industry, an entire world completely ineluctable to a stranger. In our time, patients spend billions on medicines they hope will rid them of all sicknesses. Few people reach the age of forty or fifty without having to carry pills with them. The famous physician Evgeny Tareev once said, "Most people think that a pill is a sniper bullet that will never miss its target. In reality it is more of a fragmentation grenade that is fired at an area."

My method of delivering medicine into the brain through the nasal tract aims to make the effect of the medication more precise and safe. Fortunately, the project did not require astronomical investment, since it did not aim to produce a new type of medicine, but to improve the injection of already known ones, making them more reliable, precise, and effective. Yet the funding could only be acquired for something that would impress both an expert and a businessman. In addition, in order to attract famous businessmen, good managers were needed.

At this point, our finances were handled by a large German firm. We had created a primary infrastructure and risen to a new level—and I was finally ready. My job could have proceeded more successfully and quickly if I had not stumbled upon many difficulties on the way.

The reader might ask, "How did you manage to part from Terghuys and the company he owned named TLB Electronic?"

This is what happened.

In October 1998, my lawyer Shoolman, representing Dr. Graf's interests, and I signed an agreement with Terghuys to pay TLB Electronic 350,000 guldens: 100,000 up front and 50,000 a year over the course of five years. In return, the company would waive all rights to the instrument and return the samples (there were five in all: I had two, and Terghuys had three).

We paid 100,000 to begin, but I did not pay the subsequent 50,000 on time, because I was attending a conference in the States at that time. I was reminded of the debt and warned that payment was due within two weeks. However, my new secretary did not show me this letter; instead, when I returned, she merely told me that I needed to call TLB Electronic immediately. It was Friday, the end of the week, and I sent them a fax, offering two of my devices worth 36,000 guldens and assuring them that I would pay the remaining 14,000 within two weeks. I did not receive an answer to my fax and assumed that the problem was solved. Not so fast....

In a few days, a court where Terghuys had filed a complaint froze my bank account. The court also ordered that our company's property be confiscated, and a big, nervous mess began. My lawyer warned me that it was necessary to pay the debt immediately, since, according to the agreement, if I didn't pay the 50,000 in time, then I would have to pay the balance of 250,000 all at once.

My lawyer Shoolman has to shoulder some of the blame, but some of it was mine as well. I did not read the agreement carefully when it said, "pay 250,000 at once and allow TLB Electronic to produce my advices and sell them until the company receives 500,000 guldens in profit." Not "or," as I believed, but "and," which changed the situation completely. Now I learned the worth of every letter in the contract.

Of course, I lost the case. I was forced to pay 250,000, plus offer Terghuys compensation to waive his right to produce and sell the device—another 12,000. The trial was in Dutch, and the judge gave me only a minute to speak—without an interpreter. It is hard to convey the humiliation and powerlessness that I experienced. In my years of living in Holland, I learned that in any lawsuit, even if

Exit from the Maze

I was hundred times in the right, I would not receive a single gulden, while TLB electronics would squeeze every penny out of me. You have to have limitless patience to achieve anything in any country. And you also have to have the naïveté and the passion of a Don Quixote.

In 2000, when all the experiments were successfully completed, Dr. Graf and I started to look for extra investors, because Dr. Graf did not want to invest anything further. I found a potential investors group. They were very interested in the project and promised to invest millions of Euros. Together, the director, manager, and I went to Frankfurt to meet Dr. Graf. Whilst meeting, Dr. Graf asked me to leave so he could "freely" continue the meeting without me. I did not understand why he had requested me to leave, but when I returned after one hour, I was told that the potential investors would give me an answer the following day in Amsterdam. The next day, they said no. They would not be investing any funds, even though they had been very interested the day before. I was very shocked to hear this, as I did not understand what could have happened for them to change their minds.

It was the first time that Dr. Graf persuaded a new investor not to invest in our project.

In 2003, Dr. Graf estimated the value of our companies and projects (ETDDS and EAG) at fifty million Euros. The letter on the following page will confirm this.

Longevity and Health

DR. KLAUS GRAF

T E L E F A X

To: Lerner Medical Technology Ltd. (B.V.)
     Attention: Dr. Lerner

Fax N°:

From:

Date: October 17, 2003

INCLUDING THIS PAGE WE TRANSMIT..... 1......PAGE(S). IN CASE OF ERROR PLEASE CONTACT US IMMEDIATELY:

Dear Dr. Lerner,

In your negotiations with potential investors please consider the following:

My cash investment in the two projects – Electroautonomograph and Intra brain Drug Delivery through the Nasal Cavity is 3.4 million Euros. This figure does not include the cost of my staff (Mrs. Weiss, Mr. Weber, Mrs. Suraweera and Mrs. Tischhauser) during the last 7 years which I estimate at a cost value being € 1.5 million. You also have to take into account your accrued compensation of € 3 million as of December 31st 2003.

The total cash investment amounts to nearly 8 million.

Considering the exceptional progress we have made during the last 6 years, i.e. patents, positive experiment results and presentations in various International symposiums in Europe and the U.S. the value of the Lerner Group actually should not be below € 50 million.

Sincerely yours

(Dr. Graf)

About some years ago I had contact with the assistant of Mr. Abramovich from Russia, named Ponomareva, about a potential investment in the projects. I had sent documentation regarding my work to her. For Mr Abramovich Graf was aware of this and promised that he would contact his bank in Spain, so that the bank would contact

132

Exit from the Maze

Abramovich's bank to make a deal. I never heard any result of this, and Graf told me, smiling or even laughing, that there had been no deal. To this day I still have suspicions regarding this deal.

Because Graf stopped financing my research and made it difficult for me to find new investors and maybe already sold my ideas to other investors, I have not been able to carry out my normal amount of research for the past ten years. In the few years before this happened I only participated in a small number of symposiums and was able to do some additional research in 2000. Because of the issues with Graf my project is still not finished and therefore we cannot start treating millions of patients. The only solution I can think of is to find a new investor to finance the coming two or three years of my research. Only then will we be able to start applying the technology to treatments of patients worldwide.

After several discussions Dr. Graf made clear that he would not invest more money in our companies.

I understand that soon I will have to stop the research I have spent sixty years working on. Nearly forty years were spent in the Soviet Union, and around nineteen were spent in countries in the West. Due to my experience, I can compare the different ways of working in the "gang of communists" country called the USSR, and the honest and decent West.

In Moscow I was the head of one of the biggest laboratories in the world, and one of the youngest doctors of medical science, which in the Soviet Union was the equivalent of a professor. I have obtained excellent results from my work. Due to the envy, enmity, and hatred of other doctors—even my friends—my laboratory in Moscow was destroyed. I was also fired as head of the laboratory. Luckily, with the help of the Soviet government (for whom I was a doctor), I moved from the field of medicine to research, at the Academy of Science. Later, I went to the West, hoping to continue my research. Even though I was over sixty years old, the Dutch doctors gave me a job. Earlier, I described my career in the West. And now, at the end of my life, my medical dream will not come true unless I can find some support. However, I still have some hope that together with a new investor I can finish the projects.

Chapter 11

Short Explanation of Noninvasive Drug Delivery System (ETDDS) to the Brain

Introduction

One of my inventions is the enhanced transnasal drug delivery system (ETDDS). My research to extend people's lives has been going on for more than sixty years. ETDDS can help improve people's quality of life as they live longer.

For more than thirty years, a large number of studies, mainly conducted on animals, have described the direct transport of a variety of compounds directly from the nose to the central nervous system (CNS) after intranasal administration. This new route would be a revolution in drug delivery, because nowadays many drugs targeting the human brain have great difficulties in passing the blood-brain barrier (BBB). In the past, authors like Merkus and van den Berg and others found evidence for direct transport of the drugs from the nose to the CNS, but without using iontophoresis,. Other authors, however, did not agree with their findings.

Enhanced Transnasal Drug Delivery System (ETDDS)

Instead of delivering drugs into the brain by means of enhanced transport across the blood-brain barrier, ETDDS is a method that circumvents the blood barrier and minimizes the amount of drug entering blood circulation, while enhancing drug transport into the brain. An extremely important feature of ETDDS is its noninvasiveness, meaning that the patient will

not experience any pain and/or other application-related unpleasant sensations.

Following are a few features of ETDDS:

- ETDDS for the central nervous system uses the olfactory pathway.

- The absence of the blood brain barrier and the direct connection from the nose to the brain makes the olfactory pathway an attractive route for drug delivery.

- ETDDS uses iontophoresis (a technique using a small electric charge to deliver a medicine or other chemical through the skin) as a physical enhancement technique.

Some advantages of the ETDDS system are as follows:

- **Non-invasiveness:** No needle or other invasive apparatus is needed. The active electrode is placed in the nasal cavity with the medicine dissolved in water.

- **Controlled drug delivery:** Via a little screen on the ETDDS apparatus, all the necessary information can be read and completely controlled with a button.

- **The possibility for self medication:** The ETDDS device is very simple to use and can be handled by the patients themselves.

- **Low incidence of adverse reactions:** The direct delivery of the medicine to the brain (rather than through the stomach and liver), minimizes the chance of any side effects.

- **Potentially high patient compliance and acceptance:** Because it is not painful, does not use needles, and is not complicated, people will accept the method easily.

Test Locations and Drugs Used

ETDDS was tested on rabbits in the Laboratory for Pharmacology and Toxicology (LPT) in Hamburg, under the supervision of the director, Dr. Phil Leuschner; and in Covance

Laboratories, Münster, under study director Dr. Ulrich Zuhlke. The following drugs were used for testing:

- Methylprednisolone hemisuccinate (Solumedrol, Upjohn Pharmacia)
- Methotrexate (Lederle)
- Levodopa (Sigma)
- Tacrine hydrochloride (Sigma)
- Octreotide acetate (Sunnyvale, California)

The safety of ETDDS was assessed by histopathological examination of the nasal epithelium.

ETDDS Test Results

- Transnasal iontophoretic delivery of each of the tested drugs resulted in much higher drug concentrations in the brain compared to passive (without the iotophoretics) transnasal delivery and systemic delivery.
- Applying the current for a brief time appeared to be sufficient to increase the drug levels in the brain.
- No differences in nasal tissue were observed in rabbits and monkeys treated with 1) transnasal iontophoresis, 2) passive transnasal delivery, and 3) systemic delivery. The current did not damage the nasal epithelium.
- In the experiments in Hamburg, exsanguinated rabbits were used. Because the blood was drained out, no drug transport via the bloodstream could occur. The outcome of these experiments confirmed the previous results.

Experimental Procedures

Preparation of animals:

The animals were anaesthetized with an intraperitoneal injection of urethane (Riedel de-Haen). A catheter was implanted into the arteria femoralis for blood collection. The trachea was intubated and the back of the animal's head and neck were shaved. I describe here the experiments that were performed on rabbits and monkeys, which may be disturbing or graphic for some people. Despite the seeming cruelty of the procedure, the animals were all handled as humanely as possible. The application for people will differ and will not be unpleasant.

Transnasal iontophoresis:

An Endomed 581 iotophoresis device (Enraf Nonius, Delft NL) and IOMED device was used. A medium-frequency interrupted direct current was applied to the silver/silver-chloride nasal electrodes. The electrodes with sponges containing the drug solution were installed deeply in both nostrils. The return electrode, enclosed in a sponge that was wetted with a 0.9 percent saline solution, was placed on the shaven back of the head. Silver electrodes were used during anadol drug delivery, whereas the silver-chloride electrodes were used during cathodal delivery. The following current intensities were used in the experiments: 0.0, 0.7, 1.5, and 3.0 mA. Duration of transnasal application was thirty or sixty minutes.

Histopathological examination of rabbits and monkeys' nasal mucosa was made in Covance's laboratory.

Conclusions from Test Laboratories

* * * * *

Laboratory of Pharmacology and Toxicology, Hamburg—Director Dr. Phil. J. Leuschner:
"Transnasal iontophoresis is a safe and noninvasive method to introduce drugs into the brain. Drug delivery levels can be effectively increased by means of transnasal iontophoresis."

* * * * *

Covance Laboratories Münster—Study director Ulrich Zühlke:
"On the basis of the obtained results, the safety of transnasal iontophoresis was confirmed by histopathalogical examination of the nasal mucosa. Neither any irritation nor any lesions were observed following transnasal iontophoresis when compared to controls. The

results of the five medicines that were tested showed good results. For example the test with the drug methylprednisolone showed that brain levels can be effectively increased using transnasal iontoporesis. The results obtained allowed the recommendation of clinical trials in human volunteers."

* * * * *

In order to provide objective information regarding the return on investment, I requested the University of Rotterdam to calculate these figures. Below I present a copy of the information presented to me. This financial information has not been included to show the amount of money that can be earned with this invention but to show the potential market and number of patients for which this invention can become available.

These calculations were made in 2008 by students who graduated with Master of Science degrees from Erasmus University, Rotterdam, Netherlands, on the Master of Innovation. we have analyzed the business possibilities for the enhanced transnasal drug delivery system (ETDDS), and we have arrived at the results presented below.

The investment in ETDDS will be earned back in three different ways:

1. If five million electronic devices are sold with a cost price of €50 and a selling price of €200, in total the licensing company will earn €150 x 5 million = €750 million.

2. The second channel of income will be the licenses. In total, ten licenses can be sold for €5 million each, so €50 million will be earned. The first licenses can be sold in 2011 or 2012.

3. The third channel of income will be a part of total income: €350 million from the intranasal drug delivery medicine. Plus, the company will get royalties from the net annual sales.

Short Explanation of Noninvasive Drug Delivery System (ETDDS) to the Brain

Return on Investment Examples

- Treating depression in United States: 21,100,000 patients x 365 days = 7,701,500,000 intranasal deliveries at €15 each. In total, treating depression in North and South America alone will generate €115,522,500,000 in one year. On a global scale, this amount will be two to four times higher.

- Treating Alzheimer's Disease in United States 5,690,000 patients x 365 days (minimum) = 2,076,850,000 intranasal deliveries at €15 each. In total, treating Alzheimer's disease in North and South America alone will generate a total of €31,152,750,000 in one year. For the whole world, the amount will be two to four times higher.

- Treating Parkinson's disease in the United States: 1,350,000 patients x 365 days (minimum) = 492,750,000 intranasal deliveries at €15 each. In total, treating Parkinson's disease in the United States alone will generate a total of €7,391,250,000 in one year. In the whole world, the amount will be two to four times higher.

- Treating stroke in the United States: 4,000,000 patients x 365 days = 1,460,000,000 intranasal deliveries at €15 each. In total, treating strokes in the United States alone will generate a total of €21,900,000,000 in one year. For the whole world, the amount will be two to four times higher.

- Treating epilepsy in the United States: 2,700,000 patients x 365 days = 985,500,000 intranasal deliveries at €15 each. In total, treating epilepsy in the United States alone will generate a total of €14,782,500,000 in one year. For the whole world, the amount will be two to four times higher.

- Treating drug and alcohol addiction in the United States: 25,000,000 patients x 365 days = 9,125,000,000 intranasal deliveries at €15 each. In total, treating drug and alcohol addiction in the United States alone will generate a total of €136,875,000,000 in one year. For the whole world, the amount will be two to four times higher.

- Treating schizophrenia in the United States: 2,400,000 patients x 365 days = 876,000,000 intranasal deliveries at €15 each. In total, treating schizophrenia in the United States alone will generate a total of €13,140,000,000 in one year. For the whole world, the amount will be two to four times higher.

- For CNS diseases, including the seven big CNS diseases listed above, the total income, after medicine royalties are deducted, can be around €350,000,000,000. I realize that 10 percent of this amount is more realistic, which is a minimum of €35,000,000,000.[1]

I would like to approach this data with prudence, however, so I have taken the following points into consideration.

Not every patient will use ETDDS every day for a whole year. It must also be considered that maybe not all fifty million potential patients will make use of ETDDS, plus the medication may cost less than €15. *Taking this into account—thus assuming only 10 percent of the U.S. potential is realized—then a profit of €35,000,000,000 will be achieved per year only for the U.S. and for only 7 diseases. On a global scale this amount can be 2 – 4 times higher. But there are thousands of illnesses present to humankind and many of them can be targeted by this system. Therefore this is an new epoch in medicine. As the adoption of ETDDS technology grows, so will the yearly revenue. This means that our assumption of 10 percent is a minimum for the early days of ETDDS; this percentage will increase greatly in the future.*

Sixty or seventy years ago, when scientists predicted the arrival of the nuclear bomb, it was hard for the everyday person to understand the full impact it would have on the world. Now there are thousands of nuclear bombs worldwide, and everyone knows what a nuclear bomb is, including its effect. I believe this is a good

---

[1]. National Institute on aging - National Institute of mental health - Parkinson's disease foundation www.schitzophrenia.com - American Heart Association - American Alcohol Drug Information Foundation (AADIF) - Epilepsy Foundation of America - World Health Organization www.who.int

Short Explanation of Noninvasive Drug Delivery System (ETDDS) to the Brain

comparison for the process of accepting of new technology. It may seem that 35 billion dollars (in the United States alone) is an unrealistic sum. However, in the future it will be clear that this prediction is justified, as the effect of ETDDS will be as revolutionary as the nuclear bomb once was.

Overview of Worldwide CNS Diseases from 2004–2008

Every year, millions of people get sick and die from CNS diseases, and this number keeps growing every year.

According to a 2007 report from the World Health Organization (WHO), brain disorders affect up to one billion people worldwide, a number that is set to increase as our population ages. In the United States alone, more than sixty million Americans are afflicted with a brain disorder, which amounts to nearly *one in five*.

Scientists know of more than one thousand disorders of the nervous system. They have many different causes and come in many different forms, such as neurodegenerative diseases, mental illnesses, and dementias. Some of the more familiar are Alzheimer's disease, Parkinson's disease, stroke, encephalitis, meningitis, arachnoid cysts, Huntington's disease, certain kinds of infections, multiple sclerosis, anxiety disorders, depression, schizophrenia, personality disorders, sense organ diseases, addictions (nicotine, drug, and alcohol addiction), eating disorders, learning disorders, sleep disorders, chronic pain, traumatic brain injury, spinal cord injury, and so on. A large percentage of the one billion brain disease patients have some sort of addiction like heavy smoking, drug abuse, or alcoholism. Of the roughly seven billion people currently inhabiting the globe, one billion are sick; this means that *one in seven of us* is sick. An estimated 6.8 million people die every year from brain disorders. Brain disorders do not discriminate; they affect people in all countries, irrespective of age, sex, education, or income. WHO claims that unless immediate action is taken globally, the burden of brain disorders is likely to become an even more serious threat to public health.

The 6.8 million people who died of brain disorders last year brings the total number of victims in the last ten years to more than sixty-five million, almost as many as the seventy million victims during the Second World War (WWII). In fact, the problem is so severe that one can even say that the rising incidence of brain diseases worldwide is tantamount to World War III. The table below compares the total number of victims and patients affected by ten years of brain diseases and seven years of the Second World War.

CAUSE OF DEATH OR INJURY	NUMBER OF DEATHS WORLDWIDE	NUMBER OF WOUNDED OR SICK WORLDWIDE
Brain diseases (in the last year)	6.8 million	1 billion
Brain diseases (in the last decade)	65 million	Many billions, exact number unclear
World War II (1939–1945)	70 million	240 million

The table shows that, the last year alone, the number of ill people was many times greater than the total of wounded people during all of World War II.

From these tables, it is clear that people became ill and died every year due to insufficient treatment. I hope that my methods will improve the medical treatment for these diseases.

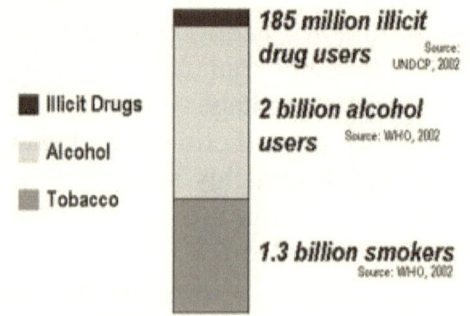

■ Illicit Drugs

Alcohol

■ Tobacco

185 million illicit drug users Source: UNDCP, 2002

2 billion alcohol users Source: WHO, 2002

1.3 billion smokers Source: WHO, 2002

The diagram above shows the potential of alcohol users to become abusers (and thus also ill). Not all alcohol users will necessarily become ill, but alcohol use does create a base on which alcoholism can form. It also shows the number of people currently suffering from tobacco and drug addiction who could potentially be treated.

Short Explanation of Noninvasive Drug Delivery System (ETDDS) to the Brain

Research has shown that every fourth young European aged fifteen to twenty-nine dies for reasons related to drinking alcohol. In Eastern Europe, alcohol is a cause of every third death among young people. Each year, approximately fifty-five thousand young Europeans die as a result of the negative effects of alcohol overuse.

The indirect costs of alcohol drinking related to social welfare, health care, insurance, administration of justice and prison management, and costs related to losses in the production sector represent a considerable portion of the total income of the community of Europe (1–3 percent). The European Commission has estimated that one-fourth of all deaths related to road accidents are caused by alcohol. The costs related to drinking and driving in the EU are approximately €40 billion a year.

The European countries are the largest consumers of alcohol worldwide, with the extent of damage, number of diseases, and early deaths caused by alcohol overuse still growing. The European Alcohol Action Plan 2000–2005, based on scientific analyses and data, recommends undertaking structured action aimed at mitigating damage in this respect. However, the research and numerical data available regarding alcohol problems are still insufficiently drawn upon in the process of developing and implementing specific strategies of preventive, educational, and therapeutic activities.[1]

Area	Annual Deaths from Alcohol-Related Causes
World	2.3 million
United States	75,000
Europe	115,000
Russia	700,000
Canada	25,000

[1]. National Institute on aging - National Institute of mental health - Parkinson's disease foundation www.schitzophrenia.com - American Heart Association - American Alcohol Drug Information Foundation (AADIF) - Epilepsy Foundation of America - World Health Organization www.who.int

Longevity and Health

It is apparent that there is an alarming rate of alcoholism and alcohol abuse worldwide. What can be done about these global statistics on alcohol abuse and alcoholism?

According to WHO statistics on alcohol abuse and alcoholism, about 140 million people throughout the world suffer from alcohol-related disorders. In the United States, alcoholism affects roughly 4 percent of the overall population, or 12.5 million men and women.

However, alcohol causes more than half of all deaths in Russia, and the problem is getting bigger and bigger. According to official statistics, in 2009 Russia had about seven million alcoholics. Experts indicate that actual national totals are a lot higher. Each Russian drinks twenty-seven liters of alcohol per year. The WHO estimates stress the point that a nation is likely to die out if its citizens drink an average of eight liters of alcohol per year.

Alcohol is now killing Russians in the prime of life at an alarming rate. Over a ten-year period, 52 percent of Russian deaths among people between the ages of fifteen and fifty-four were from alcohol-related causes. This is one of many startling facts from a study recently published in the British medical journal *The Lancet*.

The statistics of Russia's current alcohol problems are staggering:

- The average Russian drinks fifty bottles of vodka per year.

- A bottle of beer costs less than a bottle of drinking water.

- The average life expectancy for Russian men is sixty years, compared with seventy-five for American men.

- Russian women have an average life expectancy of sixty-seven, compared with eighty for American women.

- Between 600,000 and 700,000 Russians die each year from alcohol-related causes.[1]

According to statistics on alcohol abuse, in Asian countries like Japan alcohol abuse has become a major social issue. This is mainly due to the fact that drinking is a tradition when conducting business. Bars have become an extension of offices, places where major decisions are made.

---

[1]. The St. Petersburg Times, June 30, 2009. Reuters.com, August 12, 2009. World Health Organization (WHO)

Short Explanation of Noninvasive Drug Delivery System (ETDDS) to the Brain

This socially accepted way of conducting business is so entrenched in Japan that a person who declines an invitation to an after-work alcoholic drink runs the risk of being passed over for advancement or promotion within the company. As a result, alcohol is readily available in Japan in the form of vending machines that dispense cans of sake and beer.

Alcohol abuse and world statistics reveal some interesting information. For instance, as prevalent as alcohol consumption is in countries such as Japan and Russia, France has the highest rate of alcoholism in the world, however, the World Health Organization reports that binge drinking by young people is probably increasing in developing countries. Alcohol ranks at number eight in the top ten leading causes of death, killing 2.3 million people a year, compared with 5.1 million deaths from tobacco.

Around 100 million people died because of tobacco use in the twentieth century. According to the World Health Organization, almost five million people die every year from smoking-related illness.

Tobacco-related deaths annually:

438,000 America	1.2 million China	6,000 Cuba
650,000 Europe	50,000 Spain	19,000 Australia
45,000 Canada	49,000 South Korea	22,000 Saudi Arabia
300,000 Russia	50,000 Iran	86,500 England
140,000 Germany	34,000 Egypt	66,000 France
800,000 India	13,000 Scotland	6,000 Ireland

Quitting smoking and getting healthy is the top New Year's resolution every year.

The WHO estimates that there are approximately 1.1 billion regular smokers in the world, which is one-third of the global population aged fifteen years and older. Of these, about 80 percent live in low- and middle-income countries. Partly because of growth

145

in the adult population, and partly because of increased consumption, the total number of smokers is expected to reach about 1.6 billion by 2025. Worldwide, 47 percent of men and 12 percent of women smoke a total of six trillion cigarettes a year. The death toll is expected to rise to ten million per year by the 2020s or early 2030s.

Approximately 3 percent of the world population (185 million people) has abused drugs during the previous twelve months, according to the United Nations Office on Drugs and Crime (UNODC). A small percentage of the world population abuses cocaine (thirteen million people) or opiates (fifteen million abusers of heroin, morphine, and opium). By far the most widely abused substance is cannabis (used at least once a year by over 150 million people), followed by the amphetamine-type stimulants, or ATS (38 million users, among them eight million users of ecstasy).

"After the significant growth of drug abuse in the past half-century, the spread of drugs in the world has slowed down. Less than one adult person out of thirty (5 percent of the world population aged fifteen to sixty-four) has used illicit drugs in the past twelve months. The number of people that consume tobacco is seven times larger, involving a staggering 30 percent of the world population," said Antonio Maria Costa, executive director of UNODC, presenting the World Drug Report 2004 at a press conference in Moscow, hosted by Sergei Lavrov, minister of foreign affairs of the Russian Federation.[1]

---

[1]. UNODC in a two-volume World Drug Report

Short Explanation of Noninvasive Drug Delivery System (ETDDS) to the Brain

Our Ref:	Country	Status	Inventor	Title	Filing Date	Serial No.	Grant Date	Patent No.	Expiration Date
4230-2	U.S.	Granted	Lerner	APPARATUS AND METHODS FOR MEASURING AUTONOMIC NERVOUS SYSTEM FUNCTION	September 14, 2000	09/661,353	Dec. 3, 2002	6,490,480 B1	Sept. 17, 2020
4230-3	U.S.	Granted	Lerner	DEVICE FOR ENHANCED DELIVERY OF BIOLOGICALLY ACTIVE SUBSTANCES AND COMPOUNDS IN AN ORGANISM	May 20, 1998	09/077,123	Jan. 13, 2004	6,678,553	Jan. 23, 2017
4230-4	U.S.	Granted	Lerner	METHOD AND DEVICE FOR ENHANCED DELIVERY OF A BIOLOGICALLY ACTIVE AGENT THROUGH THE SPINAL SPACES INTO THE CENTRAL NERVOUS SYSTEM OF A MAMMAL	January 18, 2002	10/050,183	July 5, 2005	6,913,763 B2	Nov. 20, 2018
4230-5	U.S.	Granted	Lerner	METHOD AND APPARATUS FOR ENHANCED AND CONTROLLED DELIVERY OF A BIOLOGICALLY ACTIVE AGENT INTO THE CENTRAL NERVOUS SYSTEM OF A MAMMAL	January 18, 2002	10/051,817	Apr. 25, 2006	7,033,598 B2	Nov. 20, 2018
4230-6	U.S.	Granted	Lerner	DEVICE FOR ENHANCED DELIVERY OF BIOLOGICALLY ACTIVE SUBSTANCES AND COMPOUNDS IN AN ORGANISM	March 21, 2003	10/393,254	April 3, 2007	7,200,432 B2	Nov. 19, 2016
4230-7	U.S.	Granted	Lerner	METHOD AND DEVICE FOR ENHANCED DELIVERY OF A BIOLOGICALLY ACTIVE AGENT THROUGH THE SPINAL SPACES INTO THE CENTRAL NERVOUS SYSTEM OF A MAMMAL	October 20, 2003	10/687,816	Sept. 22, 2009	7,593,770 B2	Nov. 19, 2016
4230-8	Mexico	Granted	Lerner	DEVICE FOR ENHANCED DELIVERY OF BIOLOGICALLY ACTIVE SUBSTANCES AND COMPOUNDS IN AN ORGANISM	May 20, 1998	984007	October 6, 2006	240861	Nov. 19, 2016
4230-9	U.S.	Granted	Lerner	ADMINISTERING PHARMACEUTI-CALS TO THE MAMMALIAN CENTRAL NERVOUS SYSTEM	November 20, 1998	09/197,133	June 25, 2002	6,410,046 B1	May 20, 2018
4230-10	U.S. (Provisional)	Pending	Lerner	DEVICE AND METHODS FOR ENHANCED MULTI DELIVERY OF BIOLOGICALLY ACTIVE SUB-STANCES INTO AN ORGANISM AND TO PREVENT LOCAL IRRITATION	May 19, 2009	61/213,226			
4230-11	U.S. (Provisional)	Pending	Lerner	A SYSTEM (DEVISE AND METHOD) TO AUTOMATICALLY (QUICKLY) TREED HEART'S DISEASES	November 12, 2009	61/272,868			

These are only the US patents being described. I have patents in other countries as well.

The table above is an overview of ten of my sixteen patents on ETDDS and EAG.

The charts on the following pages show the results of the experiments with various drugs.

1. The first chart shows methylprednisolone intrabrain delivery through the nasal cavity, using iontophoresis. The columns show the concentration of the drug in the brain by delivery with iontophoresis (active). The red column shows the concentration of methylprednisolone when delivered without iontophoresis (passive). The left side of the chart shows levels in the brain; the right side of the chart shows levels the plasma (blood).

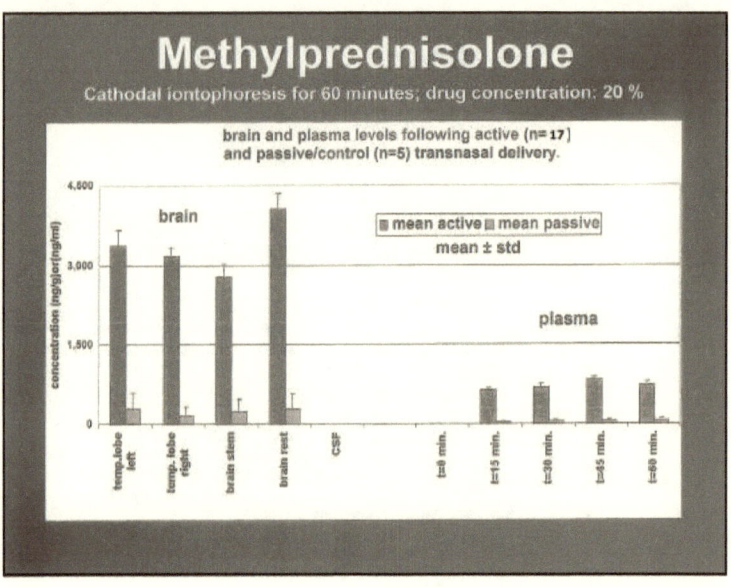

2.   The second chart differs from the first chart in that we expanded our delivery method to include delivering methylprednisolone through the nasal cavity with iontophoresis and in another group of experiments, we delivered methylprednisolone through intra-arterial

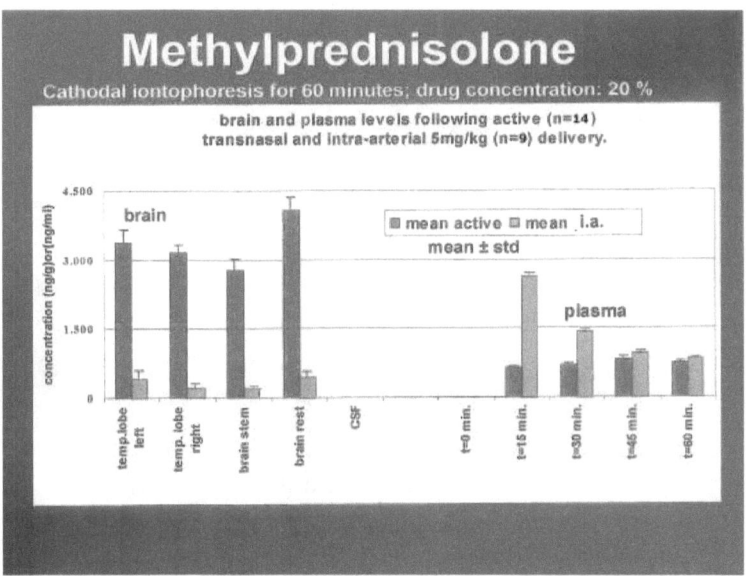

3. Methotrexate was used in the same way as methylprednisolone shown in Chart No.1 (active and passive).

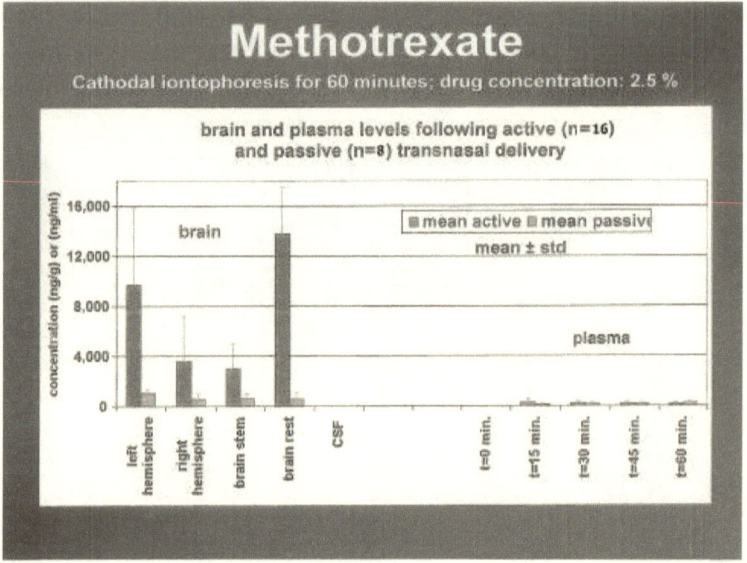

4. Methotrexate was used in the same way as methylprednisolone shown in Chart No. 2 (active and intra-arterial).

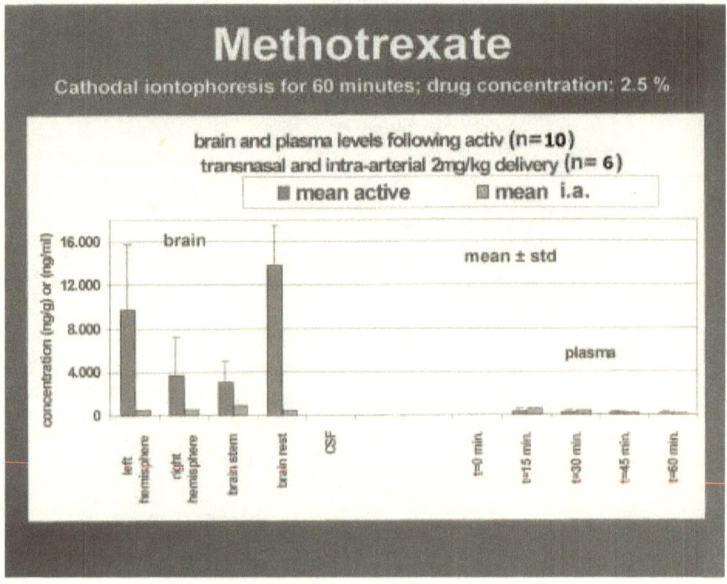

5. Levodopa was delivered in the same way as methylprednisolone, shown in Chart No.1 (active and passive).

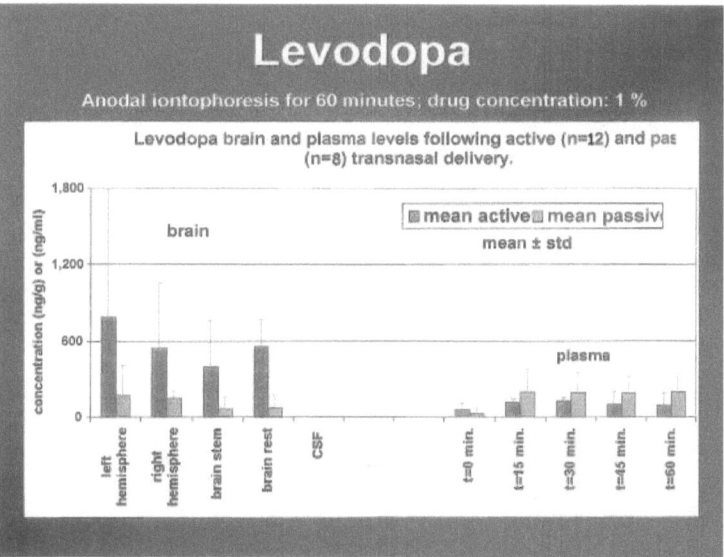

6. Tacrine was delivered into the brain through the nasal cavity using iontophoresis after the exsanguination of the rabbits (active and passive).

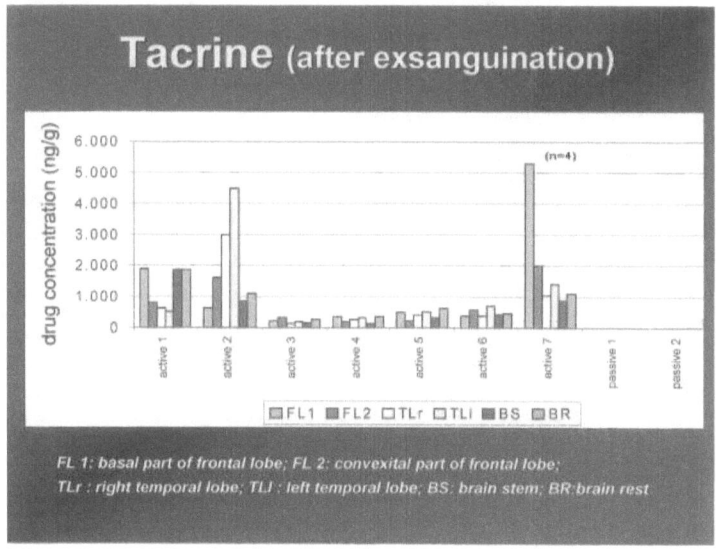

7. Octreotide was delivered into the brain after exsanguinations (active and passive).

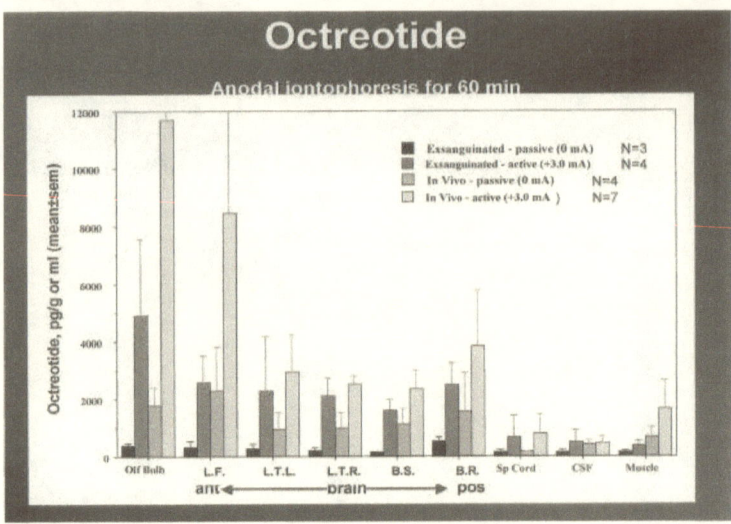

All experiments show that when drugs are delivered into the brain through the nasal cavity using iontophoresis, the concentration is much higher compared to the concentration when we delivered the drugs without iontophoresis (passive or intra-arterially). The tests performed by Dr.Ulrich Zuhlke made use only of the test medicine methylprednisolone, due to financial restrictions.

These five drugs have been chosen due to the kind of illnesses they treat—like strokes, Parkinson's disease, brain trauma, and many others—that are difficult to cure.

Short Explanation of Noninvasive Drug Delivery System (ETDDS) to the Brain

24ᴱ SALON
INTERNATIONAL
DES INVENTIONS
GENÈVE 1996

Après examen, le Jury International a décidé

de remettre à: Monsieur Eduard LERNER

pour l'invention: Système non-invasif d'administration intracérébrale
de médicaments

UNE MEDAILLE DE BRONZE _____   Genève, le 22 avril 1996

le Rapporteur du Jury

le Président du Comité
d'Organisation du Salon

This is the official statement that I won a bronze medal for my ETDDS invention.

In my laboratory in Moscow, over many years, I performed many experiments with ETDDS on rabbits and cats. The experiments were performed more than thirty times with positive results, as I have already described in this book. Over the course of many years, I treated patients who had central nervous system illnesses with ETDDS. On patients, I used ionthophorasis with 0.7 to 1.0 mA for twenty minutes, with different kinds of medicine. The treatment was performed every other day, so the total was six to ten times. The treatment resulted in much healthier patients, but these treatments were real-life treatments, never official tests or experiments. At the time, I had no equipment to check the results, but I never witnessed any negative results in patients.

On patients, I used the following drugs:

1. Suprastin
2. Promethazine
3. Nicotinic acid
4. Diphenhydramine
5. Diazepam
6. Chlorpromazine
7. Procaine
8. Vitamin B1
9. Calcium chloride
10. Acetylsalicylic (aspirin)
11. Gamma aminobutyric acid
12. Glutamic acid
13. Euphyllin
14. Ephedrine
15. Atropine
16. Hydrocortidonr hemisuccinate
17. Vitamin B12 (Cyanocobalamine)
18. Cerebrolysin

In my hospital in Moscow, I used these drugs for many patients suffering from brain diseases. I was especially busy with stroke patients. I treated many of these patients with intranasal drug delivery into the brain with methotrexate and methylprednisolone.

Short Explanation of Noninvasive Drug Delivery System (ETDDS) to the Brain

For example, U.S. Senator Edward Kennedy was ill with brain cancer. An operation removed a part of the cancer, but after some time, he died. It is known that when treating cancer in the brain, methotrexate is used. However, when this drug is given orally or through injections, most of it goes into the bloodstream, and hardly anything reaches the brain. If the doctors who had been treating Edward Kennedy, had been using methotrexate and methylprednisilone and my invention—intrabrain drug delivery through the nasal cavity, before and after the operation—Edward Kennedy would not have died, or he might have lived longer.

Many drugs can be delivered intrabrain through the nasal cavity by means of iontophoresis. Although some drugs cannot normally be delivered by means of iontophoresis, there are methods that make it possible to do so anyway. For example, by using special buffers and making the pH less than 5. Such a method makes it possible to deliver these drugs using iontophoresis as well. While not all drugs can currently be delivered with iontophoresis, the pharmaceutical companies can alter the drugs so that they can. It is a matter of preparing the drugs so they can be used in the iontophoresis apparatus.

Chapter 12

Short Explanation of
Electroautonomography (EAG)

Electroautonomography is a diagnostic tool for disorders of the autonomic nervous system

Introduction

The autonomic nervous system (ANS) is a complex system that regulates smooth muscle, cardiac muscle, and every other visceral organ (large organ) in the human body. The ANS is divided into two functionally and anatomically distinct parts: the sympathetic part and the parasympathetic part. The sympathetic division has thoracolumbar outflow, meaning that the neurons begin at the thoracic and lumbar (T1-L2) portions of the spinal cord. The parasympathetic division has craniosacral "outflow," meaning that the neurons begin at the cranial nerves and sacral spinal cord.

Disorders or injuries of the ANS can be located peripherally (PNS), consisting of the nerves and ganglia outside of the brain and the spinal cord and/or centrally (CNS, brain and spinal cord). Until now there was no method of directly measuring the ANS function. Indirect methods like ECG (R-R top) variability and evoked sympathetic skin response (showing only the status of peripheral sympathetic fibers) are nowadays the most important available diagnostic tools.

Introduction to Electroautonomography (EAG)

Electroatonomography is a noninvasive method for diagnosing disorders of the autonomic nervous system, based on the registration and analysis of skin potentials and other electrical signals. By recording spontaneous and evoked skin potentials simultaneously from both hands and feet in combination with the registration of the electrical signals of various organs,electroautonomography allows diagnosis of both the peripheral sympathetic and parasympathetic parts and reveals the extent of damage to either part. (ECG with R-R top variability, respiration, and electrogastrography: the electrical signals that travel through the stomach muscles and control the muscles' contractions) Currently, neurologists use series of clinical tests to assess he activity of the autonomic nervous system. An EAG device was tested at several universities with a positive result. The EAG must be seen as a new method to assess the ANS activity and can be used instead of all the existing methods. Because of its diagnostic power, EAG has the potential to become a very important diagnostic tool for neurologists and other physicians.

Short Explanation of Electroautonomography (EAG)

UNION OF SOVIET SOCIALIST REPUBLICS

STATE COMMITTEE OF USSR FOR INVENTIONS AND
DISCOVERIES

CERTIFICATE OF INVENTION
No: 1362445

Upon the Authority of the Government of USSR, the State Committee of USSR
for Inventions and Discoveries has issued this Certificate of Invention:
**"Method for Determining the Condition of the Autonomic Nervous System by
E.N. Lerner"**

Inventor(s): **Lerner, Eduard Naumovich**

Submitted: himself

Application No: **4224307**

Priority Date for Invention: **March 11, 1987.**

Registered in the State Inventions Archives of USSR: **September 1, 1987**

This Certificate of Invention covers the entire territory of the Soviet Union.

Chairman of the Committee: **signed**

Chairman of the Department: **signed**

Above is the translation of the Russian innovation certificate.

Expected Applications of the EAG Device

Many ANS diseases and injuries can be detected with the EAG. Neurologists will use EAG for diagnosis and disease monitoring. The diagnostic power of the EAG will make obsolete other time-consuming and often patient-unfriendly methods to assess ANS functions. EAG can be used to predict the development of a number of serious diseases many years in advance, even from childhood. Recently, for the first time in the history of veterinary science, the function of the ANS of a horse was determined using the EAG device.

This allows completely new ways of diagnosis in veterinary medicine and can actually save the lives of many horses suffering from ANS disorders that are currently misdiagnosed or undiagnosed. Pharmaceutical researchers could use the EAG as the tool to directly investigate the dose response relations of new neuro-active drugs.

The eight-channel EAG device that I developed has the ability to:

- improve calculations of autonomic nerve conduction velocity in the proximal and distal parts of the extremities.

- register four channels EAG combined with electrogastrography (EGG) electrocardipgraphy (ECG) and respiration or saturation.

- be adapted for electro-encephalography (EEG) and electro-oculography (EOG).

- calculate the heart rate variability, respiration rate, amplitude, and variability.

- perform fast fourier transformation on recorded signals.

- calculate the first derivatives of recorded signals with included software in the apparatus.

- export data for further analysis.

Market Analysis

The market for the EAG device is large, because not only neurologists but also other specialists and general practitioners can use it as a standard diagnostic tool in their daily practice. The worldwide number of potential buyers is three million. Since a prototype is already available and FDA approval will be not difficult as the EAG device is completely noninvasive and safe, we expect that market introduction could occur as early as the end of 2011. The EAG device is a completely new method and no direct competition exists, and an estimated price of €30,000 per device is realistic.

The EAG project is a unique opportunity to open new ways of diagnosis and a better understanding of the function and dysfunction of the autonomic nervous system. The impact on medical society is not easy to estimate but will be huge, since the possibilities of the EAG device are numerous and the competition is scarce. Furthermore, the possible end-users of the EAG can be found throughout the medical profession, from neurologists in academic hospitals to general practitioners. The potential usage is also high, since every disorder that includes ANS damage (central and/or peripheral) could be investigated with the EAG device. In conclusion, the EAG can become a standard diagnostic and prognostic tool for physicians all over the world.

Scientific Background—The Autonomic Nervous System[1]

The autonomic nervous system (ANS) is concerned with the regulation of smooth muscles, cardiac muscles, and every other visceral organ and tissue in the body. The system is not directly accessible to voluntary control. Instead, it operates in an automatic fashion on the basis of autonomic reflexes and central control.

The ANS is a complex system. One of its major functions is the maintenance of homeostasis within the body. The ANS further plays an adaptive role in the interaction of the organism with its surroundings.

Because the ANS is responsible for regulation of every organ, ANS-related disorders are very common. According to Kellner[1] up to 40 percent of the population suffers from some kind of ANS-related disorder. This can range from mild to severe and is sometimes life threatening since the ANS controls vital functions. A range of different clinical tests—currently estimated at more than 350—can assess ANS function.

The tests mostly show indirectly the functioning of the autonomic nervous system. Also, the tests are often laborious, are unpleasant for the patient, and the outcome is not always reliable. Electroautonomography is a noninvasive diagnostic method for direct measurement of the functioning of the ANS, based on the registration and analysis of skin potential and other electrical signals. Skin potentials originate mainly from sweat gland activity. The existence of skin potentials was first noted by Tarchanoff at the end of the nineteenth century.

Measurements of spontaneous skin potential could have been used as a direct method for assessment of ANS function, but this was

---

[1]. (Neurosis in general practice; BR J Clin Pract 1965 vol 19 (12) pp 681–2),

limited due to the instability of the signals, the high sensitivity to noise, and the occurrence of habituation resulting in the disturbance of signals. The electroautonomograph however, opens new ways for the practical use of skin potential recording, because the device uses a unique, highly sensitive DC amplifier instead of an AC amplifier, resulting in more accurate signals. With the help of Twente Medical Systems (TMSI), I improved the original EAG model into a perfect working apparatus.

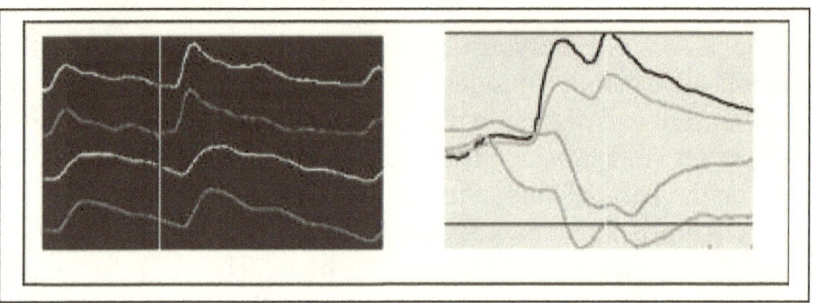

Typical EAG recordings: four channels of spontaneous skin potentials were recorded from both hands and feet. The left image shows a healthy person, while the right shows a person with chronic fatigue syndrome (CFS). The contraphase between the hands and feet is an indication of central autonomic dysregulation.

More Expected Applications of the EAG Device

EAG as a diagnostic tool

Many ANS diseases and injuries can be detected with the EAG. Using EAG, it is possible to discriminate between lesions in the central and peripheral part of the ANS.

EAG will complement monitoring systems like EEG, ECC, and EMG. Eventually, the diagnostic power of the EAG will make obsolete currently used, time-consuming, and often patient-unfriendly methods to assess ANS functions obsolete.

EAG as a prognostic tool

EAG can be used to predict the development of a number of serious diseases many years in advance, even from childhood.

Short Explanation of Electroautonomography (EAG)

A misbalance of the ANS system resulting in sympathetic overactivity or prevalence is associated with the future occurrence of hypertension-related diseases such as brain and heart infarcts. A misbalance resulting in parasympathetic prevalence is associated with the high incidence of stomach diseases, allergies, asthma, bronchialis, and more in the future. Also, ANS damage as a result from addiction to alcohol or drugs (narcomania) and immunologie disturbances can be determined. The level of stress sensitivity of a person can be predicted as well. Beside the endocrinological effect of this diseases such as asthma, allergies etc, it impacts the ANS in several ways. For example, 60–70 percent of patients develop a mild to severe form of diabetic neuropathy as a complication, which in the severe forms can lead to lower limb amputations. According to the diabetic foundation, each year more than 75,000 amputations are performed among American diabetics. Furthermore diabetic patients have a two to four times increased risk of developing heart disease; 75 percent of diabetes-related deaths are due to heart disease. As with many diseases, early detection of this neuropathy will lead to an early treatment and hence a higher incidence of therapeutic success. EAG recording will reveal in an early stage the status of peripheral and/or central part of the ANS. This property of EAG is unique. Insurance companies could include an EAG test prior to covering, determining the health status of persons with special cases and thus predicting the long-term risks.

EAG and job selection

The state of the autonomic nervous system of applicants for professions that require perfect health—such as astronauts, airline pilots, scuba divers, and many other professions—can be determined very accurately; hence, the EAG device can be used to make better selections.

EAG and veterinary selection

As in humans, animals also possess a parasympathetic and sympathetic part of the ANS. Horses that have a sympathetic overactivity can be winners in horse races. On the other hand, cows that have a parasympathetic prevalence will be great milk producers. For the first time in the history of veterinary science a horse's ANS function was determined recently by using the EAG. This opens the door for completely new ways of diagnosis in veterinary medicine, and can actually save the lives of many horses suffering from ANS disorders that are currently misdiagnosed or undiagnosed, like grass sickness, which kills thousand of horses each year.

EAG as a pharmaco-dynamic tool

Pharmaceutical researchers could use the EAG tool to directly measure the dose-response relations of new neuro-active drugs. In this way, expensive and time-consuming clinical testing can be limited, also reducing the volunteers' risk.

Après examen, le Jury International a décidé

de remettre à: Monsieur Eduard LERNER

pour l'invention: ELECTROAUTONOMOGRAPHE - Instrument servant à mesure l'état du système nerveux autonomique

UNE MEDAILLE D'OR _____ Genève, le 22 avril 1996

le Rapporteur du jury

le Président du Comité d'Organisation du Salon

This is the official statement that I won a gold medal for my EAG invention.

Market Expectation

The registration of skin potentials on the EAG will give cardiologists a tool that will provide more complete information about the heart, because the heart rate is regulated by the ANS. Any information about the status of the ANS can be of help for the cardiologist. For example, 4 percent of the healthy and young population (between eighteen and twenty-six years) have a hypersympathetic prevalence, a condition that is associated with a high chance of developing hypertension-related diseases such as cerebral vascular accidents (still the worldwide number one cause of death) or renal failure. The autonomic prevalence can be determined by an EAG recording. Early recognition of a potential risk is possible using the EAG device, thus making early treatment possible. It would also be possible to undertake adequate actions before the disease manifests itself or even to prevent the disease by changing the way of living.

Another area in which the EAG device can be an important tool for better diagnosis and prognosis is complications of the ANS in diabetes. Mellitus diabetes is the seventh leading cause of death in the United States.

Disorder	EAG Device usage	Estimered number of sufferers in USA	Potential total number of EAG tests	Estimated number of sufferers in Europe	Potential total number of EAG tests Europe
Multiple system atrophy	Diagnosis + monitoring every 2 year = 10 x	70.000	700.000	120.000	1.200.000
Parkinson's disease	Diagnosis + monitoring every 3 year = 5 x	1.200.000	6.000.000	2.000.000	10.000.000
Syncope and presyncope	Diagnosis + monitoring every 3 year = 5 x	350.000	1.750.000	700.000	3.500.000
Multiple sclerosis	Diagnosis + monitoring every 3 year = 5 x	400.000	2.000.000	380.000	1.900.000
Amyotrophic lateral sclerosis	Diagnosis only = 1 x	16.000	16.000	35.000	35.000
Brain and spinal cord tumors	Diagnosis + monitoring = 2 x	110.000	220.000	230.000	460.000
Amyloid (carpal tunnel syndr.)	Diagnosis + monitoring (2x) = 3 x	165.000	495.000	350.000	1.050.000
Diabetic autonomic neuropathy	Diagnosis + monitoring every 2 year = 10 x	1.020.000	10.200.000	1.900.000	19.000.000
Total		3.331.000	21.381.000	5.715.000	37.145.000

Potential EAG usage in diagnosis and monitoring of several ANS disorders. Based on the life-expectancy and progression of the disease, patients need to be monitoring regularly to be able to recognize and possibly prevent or treat complications or deterioration of the disease.

As with many diseases, early detection of this neuropathy will lead to early treatment, and hence a higher incidence of therapeutic success. EAG recording will reveal in an early stage the status of peripheral and/or central part of the ANS. This property of EAG is unique.

Up until now, there has been no one other method or device capable of such early detection. Finally, the diagnostic power of the EAG makes it very suitable for the office of general practitioners. Using all parameters, EAG can give the GP information that normally requires consulting a specialist. It gives the GP more possibilities to diagnose and adequately treat patients with ANS-related disorders.

Market Volume

The outlined market expectations gives us the opportunity to predict the number of EAG devices that could be sold in the countries where the EAG patents have been filed. Currently, EAG patents have been filed in the United States, the European Union, and Japan. The figure below show the number of registered physicians (specialists and general practitioners) who might use EAG.

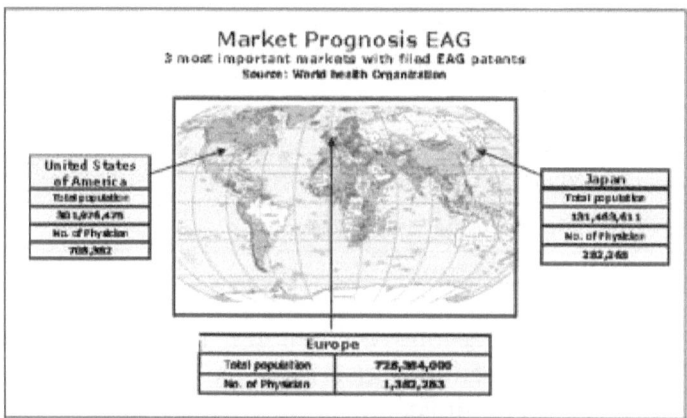

Global Market

The medical technology industry is one of the most innovative industries in the world. Built on expertise and creativity and driven by demographic trends, the $260 billion global market for medical technology is also one of the fastest growing in the world. Several factors are likely to speed up further growth of health care technology. First, an increasing elderly population in both industrialised and developing countries will greatly increase the demand for medical devices.

Pricing Options EAG

If the following options are considered for the EAG device, one can make a reasonable price estimation.

The EAG device is a truly multi-parameter device that, with special software, can be used to measure many different parameters.

EAG has the potential to become the most important device used for investigation of the ANS.

Competitor prices are in the range of €10,000–€65,000, although there is no such apparatus that covers all areas of diagnosis.

The estimated average charge for each EAG session, according to the CTP codes used by U.S. health insurance companies, could be as high as €2,000. To make the EAG method less expensive than current methods, a price of €500–1000 per test will be ideal.[1]

Clinical Test	CTP Code	Charge in U.S. $
Cardio-Vagal Changes (Autonomic Function Test)	95921	230
Blood Pressure Monitoring	95922	280
(24 Hour) Heart Rate Variability	93224	380
Electrogastrography	94834	400
Nerve Conduction Velocity	95900	500
Sudomotor Changes in Sympathetic Skin Response	95923	200
	Total	**$ 1,990**

Overall, also taking into account the prices of competitors—although no direct competition exists since there is no other device that can do the same—it is estimated that the price for an EAG device plus computer and software can be one-tenth of the price of equipment currently needed to produce diagnoses.

This price is low compared to the current prices of other devices, because, as explained, a doctor or hospital currently needs to buy several different devices to be able to make a proper investigation. It is

---

[1]. Source: Letter from H, Rashed, director of Autonomic Function Unit, University of Tennessee.

estimated that for a hospital to buy such equipment, costs are on average more than €100,000.

Investor Information

This business plan has provided you with background information about Lerner Medical Technology Ltd. and the EAG project. We hope it is clear that the EAG has tremendous potential to become a worldwide important diagnostic and prognostic tool for ANS disorders.

A number of factors support this perspective:

- There is currently no competition.

- Not only neurologists but also other specialists and general practitioners could benefit from using the EAG device.

- The reimbursement for using the EAG device could be up to $2,000 per test in the USA

What Investors Can Expect

Depending on the scenario, the return-on investment point will be reached either the moment that the agreement for selling the entire project has been made or when sales of the device become higher than the calculated break-even point (for a price of €10,000 per device, this would be sales of two hundred devices). Apart from the return-on-investment, additional terms can be discussed in a later stage.

To give an estimation of the total amount that can be expected in return, we give two examples of different markets: the Netherlands and the United States,

- In the Netherlands, there are 1,080 neurologists and 2,164 psychiatrists, constituting the most important clients of the EAG. If only 50 percent of them buy the EAG apparatus, it would generate a total of: 3,244 x 50% = 1,622 x 10,000 (selling price of EAG) = €16,220,000.

- In the United States, there are 11,200 neurologists and 39,000 psychiatrists, constituting the most important clients of the EAG. If only 50 percent of them buy the EAG apparatus, it would generate a total of: 50,200 x 50% = 25,100 x 10,000 (selling price of EAG) = $251,000,000.

Chapter 13

Notes on Medicine

My wife's misfortune and the fact that she had suffered more than needed gave me an insight into the flaws of the Dutch medical system, as displayed by one clinic. I think that these flaws were local and specific to that hospital only. What I observed in the biggest hospitals in the West leads me to this conclusion. Doctors in Holland are capable of thinking. Without question, doctors here receive good training and good salaries, but their thinking is too standardized The main principles of the old Russian medical school was to treat the patient, not the sickness—that is, to try to consider the patient's individuality. This approach was used by all the best Russian clinicians. The history of Russian medicine is peppered with stories of many doctors who, through their humanism and self-denial, gained such popular admiration and love that they were treated like saints. In 1853, when Dr. Feodor Gaas died, ten thousand mourners followed his coffin to the cemetery. On the day of the funeral of the outstanding ophthalmologist Leonard Girshman, all factories and schools were closed in the City of Kharkov. In Kiev in 1928, when the famous doctor Feofil Yanovsky died, Christian and Catholic priests showed up next to the rabbi.

In the West, the doctors are tamed, rather than nourished; they are drilled to develop a rigid cause-and-effect mode of thinking, which rules out a general view of the patient and an ability to find non-trivial solutions. If a person is diagnosed with condition A, he must be treated for A; if his sickness is B, he must be treated for B, if he is sick with C, and so on. What happens, however, when a disease does not fit types A or B and is furthermore complicated with symptoms of C? A doctor then falls into a state of confusion because of the unknown condition! Actually, the condition is well-known, it is just that it is not taught in medical school.

Longevity and Health

If devices do not show any deviations and the computer has not sounded an alarm, the doctor is calm. The patient could be complaining of stomach pains, but a doctor will not even listen. He will not ask when the pains appeared, how intensive they are, or under what circumstances do they recur, he will automatically send the patient to get an ultrasound and gastroscopy.

If a person is diagnosed with a growth or an ulcer, he is sent to a surgeon. If he finds nothing, then the patient is not ill. Such conditions as gastritis, colitis, spasms, and other "small" disorders that were treated or at least relieved in the past in Russia, are disregarded in the West. If a patient is not running a fever, then, according to local doctors, a person is completely healthy. Small respiratory infections are considered a quasi-normal condition, though as a neurologist I know that they sometimes lead to the gravest complications.

I have always been convinced that mechanical procedures are of great importance (and this opinion has led to problems with other doctors, who criticized my excessive use of "hardware"), but clinical mentality is just as important. A few particular incidents in my practice in Holland confirm this conclusion. On my first visit, in 1988, I met with the Marsen family. They told me of a young American girl named Susan Wilner, twenty-eight, a gynecologist who suffered from a rare disease only described in medical literature. Patients with this condition cannot eat salty foods, or else they will swell, their heart will fail, and they will die.

While traveling in Holland, Susan carelessly ate something salty; she realized it in time and rushed to the hospital immediately, where she told the doctors her problem and asked for medication. Instead of rendering emergency treatment, the doctors discussed her condition. She lost consciousness and her heart stopped. Fortunately, Susan was still in a hospital. Her heart was restarted, and she was placed in critical care.

I was told about it ten days after it happened, when Susan's parents had flown in from America. Since the Marsens knew I was a neurologist, they asked me to pretend to be Susan's relative and visit her to check on her condition. I didn't want to do it, but they managed to convince me. There were no doctors on duty in sight, so we walked into her room. Susan was in 1 horrible condition. She was almost unconscious, had continuous cramps, and was suffering a tonic convulsion that means that

172

she suffered from a special kind of spasm. I asked the nurse about the patient's treatment. She told me that an IV had been placed, a combination of vitamins was given, hormones and mineral salts.

"What else?" I asked.

"Nothing." she answered.

When I left the room, her parents bombarded me with questions, but how could I decide whether their daughter was getting the appropriate treatment without knowing her medical history? Susan's parents confessed to me that Susan had been seen by a doctor from Moscow. The local doctors were curious to know my opinion and invited me to the hospital. I met the general practitioner and neurologist, who confirmed that they had prescribed the mix of vitamins, hormones, and salts to keep her alive. The computer tomography did not show any pronounced signs of brain damage, and they were at a loss what to do.

I was surprised, because to me, the picture was clear. As a result of the heart stoppage, she experienced hypoxia (lack of oxygen) of the brain. It was natural that the computer showed no changes, since she had no local damage like hemorrhage, softening, et cetera. I had run into similar cases and treated them successfully, and I told the doctors what I thought.

They listened attentively. Then the professor invited me to the computer to tell him what I had made of the received results. I confessed that my hospital had no such equipment, and therefore I could not make any comments. His attitude toward me changed immediately. I even noticed grins in the faces of the professor and the doctors. After all, what could there be to talk about to a Russian who knows nothing about computer and tomography? Our conversation ended with them making the most pessimistic prognosis, the patient is dying, and there is nothing that can be done and whatever I proposed would be completely useless.

On the following day I flew to Moscow. When I returned to Holland six months later, I called up the Marsens to find out about Susan. It turned out that she had survived. Despite their skepticism and smirks, the doctors decided to follow my recommendations. After a few hours of intensive treatment, the cramps stopped and her condition drastically improved. She was able to recover, though she ended up disabled.

Longevity and Health

A year later, I suddenly got a call from Mrs. Marsen to tell me they received a visit from the Wilner family: Susan, her mother, and her brother, who turned out to be a psychiatrist. Mrs. Marsen invited me over. Susan opened the door herself, the woman whom I had seen only in the hospital. She had returned to the United States completely disabled, with contracture (tightness of joints) in her limbs. After a long search she found a rehabilitation center in Germany, where she had a course in treatment and got rid of the contracture. Though she still exhibited a slightly higher tonus in her limbs (spasms), she could move around by herself and use her hands almost like a healthy person.

A remaining complication was her optic agnosy, a corruption of cognitive function where one remains lucid and able to see without understanding what they are seeing, caused by damage to the cortex. If the treatment had been applied immediately, these complications could have been avoided.

For ten days she was in the hospital unattended by doctors. This would not occur in a Russian village, but it did in the center of Europe, in a wonderfully equipped hospital in Amsterdam. The problem was the doctors' over-reliance on computers to understand what happened, and this almost cost Susan her life. In my medical practice, I have had many such patients. I have always been able to help them, especially the young people who had sufficient compensatory mechanisms to overcome their condition.

I have made many complex diagnoses, but I want to mention just one. About 150 years ago, Professor Kozhevnikov diagnosed two brain tumors, on the basis of clinical symptoms alone. Thus, he made his name as an outstanding clinician and the founder of Russian neurology. "I found that I have done the same thing," but I found I have been capable of doing that very same thing.

A very sick patient was brought to my department at the end of the day on a Friday. We could do detailed examination on Monday, but we had to diagnose a case as severe as his right away. I examined him (with two other doctors present) and diagnosed two tumors: one in the brain, and the other one in spinal marrow, which probably would lead to death. Neither of the two doctors believed the diagnosis.

The next day, the patient died. Many doctors and members of the administration came to the meeting afterwards to discuss the case. They tried to get me to change the diagnosis and argued that Kozhevnikov's methods were antiquated. Yet I persisted, and when the autopsy confirmed my diagnosis, all those present went into a stupor. Objectively, from the diagnostic point of view, the case was unique.

Yes, Dr. Feodor Gaas was right a thousand times when he said, "Medicine is the hardest science. Not because there are so many diseases or because it requires the knowledge of so many other subjects, but because not a single element of any of its problems can be precisely calculated. There is only a close estimate. Only a brilliant doctor can do it, one helped by what I call the sense of experience, one of the most refined human qualities."

Of course, even good doctors make mistakes. Once I developed pains in my lower back. It happened on a Friday, but with the help of an experienced and trustworthy doctor in Utrecht, I was able to get an MRI. However, the radiologist was already gone and there was no one else to read the results. The doctor who had helped me get the test went to his office to take a look at the results on the computer. Suddenly he exclaimed, "Oh! This is cancer with metastasis! Look, here it gets into lymph nodes, and here's metastasis in the bones."

The following weekend became the blackest of my life. I barely was able to ask, "So, how much time do I have left?"

"You've got a while," he said, "if we start you on steroids and hormones right away."

In other words, a year or two.

On Monday, at ten in the morning, he called up to tell me that the MRI radiologist didn't think I had either cancer or a metastasis and that everything was absolutely normal.

In another case, I began to have uncomfortable sensations in my heart. A doctor from the city of Haarlem did an ECG.

"You have had a vast heart attack! Don't move!"

He called an ambulance, and I was taken out on a stretcher. The cardiologist at the hospital laughed when he saw the cardiogram. By accident the doctor had switched poles, putting a plus instead of a minus and vice versa.

A Doctor, I knew well and trusted, told me a story. He had a patient who had suffered a cranial injury, and then later neurological and internal symptoms. He went to see a number of doctors with different specialties.

Longevity and Health

His visit to the neurologist lasted a mere couple of minutes. The doctor asked when the injury had taken place and, without examining or asking about complaints, told the patient to go home, with no advice or prescriptions.

The patient went to see a gastroenterologist. Again, without examining or asking about complaints, the doctor sent the patient for a gastroscopy, which was done by a young, female gastroenterologist. The procedure took about a minute or two. The patient was told there was no pathology and there was no reason to go back to the gastroenterologist who had ordered the procedure. Next was the ENT specialist. Again, a minute or two, without an examination or asking about complaints.

Due to the heavy cranial injury, the family doctor sent the patient to obtain an MRI. The patient asked the woman at the MRI office if she was a doctor. The woman, who was a nurse, told him that if he was not happy, he didn't have to go through with this. During the procedure she turned his head aside, which is a definite no-no, thus she messed up the procedure and categorically refused to do it over.

I have practiced medicine for over sixty years, and have been a doctor of science for over forty-seven, and I have spent nineteen of these years in Holland. I can cite a tremendous number of cases of irresponsible and downright criminal attitudes by doctors toward their patients. The problem is especially widespread among Russian doctors. Many of them emigrated to the West, mostly to the United States. By hook or by crook, they passed the boards and now they "treat" patients. Many of them actually present a danger to their patients. If local doctors behave irresponsibly, as I pointed out earlier, the ones from Russia should be avoided. Being a Russian-born doctor myself, I take no pleasure in saying this, but years of experience have convinced me.

As an example, one relatively young man that I knew from Russia, who was forty years old, emigrated from Russia to Holland. He became very ill, and the Dutch doctors diagnosed him with lung cancer. In their opinion it was too late to operate on him, but radiation therapy would extend his life by several years. He trusted Russian doctors, so he went to Moscow.

The surgeon in Moscow promised him that an operation would cure him, so he agreed to pay the large sum of money required for the operation. Sadly, he died shortly after the operation. Because the cancer was not very aggressive, he could have lived for several years with radiation therapy. From this you can see how the biggest Russian surgeons conduct business. Sometimes they are more interested in making money then treating the patient. It is not that I am against making a profit, which I also need, but my first interest is helping the patients.

Many American doctors also think only about profit. They send patients for many tests and consultations with other doctors who will, in return, send them new patients. An example of this was when I was in San Francisco and a friend of mine had an appointment with a famous cardiologist. After the doctor did the preliminary examination, he sent my friend to his nurse to complete the examination. The nurse gave him many forms for appointments for many tests. When the doctor saw what the nurse had done he told her that in this case, because my friend was a colleague and a doctor, it was not necessary to do many tests, only to send him for an EKG.

In Europe, the doctors in hospitals have a fixed salary. They usually try to examine a patient very quickly and then send them for one or two tests, to be performed in one or two months, and then the patient will return to the original doctor in three or four months. If during this time the patient dies, the doctor is not held responsible.

Having visited the largest medical centers in Holland, Germany, Belgium, England, the United States, and other countries, I became convinced that medicine is at a relatively low level. It cannot even be called a science. While we see some progress in surgery, general medicine lags behind. In other words, we succeed where *hands* are involved, but lag behind where we use the *mind*.

Modern Russian medicine is a combination of science, guesswork, and direct dishonesty to the patients, from whom doctors can get more money. This criticism does not apply to doctors working before Stalin's terrorism, who were highly educated. In the West, doctors never cheat but don't want to show their ignorance. This is because their medicine is not a real science. Therefore, Western doctors should not be trusted very much, and one should always get a second opinion.

Modern medicine's weakness is caused by two things. Firstly, medicine is one of the most complicated sciences. Secondly, governments do not give enough financial support for the science.

Private people invest money in practical medicine only, and not in research, because they can make a profit. Nowadays, my personal goal is combining medical research and making a profit

On the whole, my contact with doctors and nurses has convinced me of the relatively high level of Dutch medicine. Of course, it would be ridiculous to deny that progress comes from the West and not the East. The list of Nobel Prize winners in medicine in the last twenty years includes scientists from the United States, France, Canada, Switzerland, Germany, Great Britain, Italy, and Denmark. The creation of psychopharmacology that led a revolution in psychiatry; the application of antibiotics, sulphate, steroids, and hormones, transplants; operations on the heart and vessels; the use of fiber optics, laser medicine, computer tomography, gene engineering; and finally the vast effort to map the human genome—all originated in the West.

In my memory, not one major discovery in the field of medicine has been made in the former Soviet Union in the last fifty years. If I ever become sick, I would without hesitation prefer to be treated in Holland instead of Russia. Perhaps, considering what my wife had to go through, this sounds paradoxical. But the overall level of Western medicine, especially in surgery, is immeasurably higher here than in Russia. In addition, the West provides excellent conditions for the patients, fully equipped hospitals, and fully qualified medical staff. Here you cannot even imagine that a hospital would not have hot water, heat, bed sheets, bathrooms for the disabled, hospital cots, and enough surgical tools, medicine, bandages, and other equipment. Doctors and patients in Russia have to do without many of these things.

You can't trust doctors in Russia. Once, I had pains in my stomach, more on the left side. My wife suggested that I consult with a surgeon from her institute (MONIKI). I drove to see the surgeon and walked into an operation room. He reeked of alcohol as he lay me down on the table and leaned over to feel my stomach. He said that I had acute appendicitis, and I needed to be operated on immediately.

"My pain is mostly on the left," I said. And of course, the appendix is on the right side I told him.

"Sometimes it radiates to the left," he countered.

I begged him to let me go for a few minutes to call my wife. As soon as I stepped out of the room, I jumped into my car and drove home. On the following day I discovered that I had kidney pains. I was fortunate enough to avoid surgery. This is just another example of Russian medicine.

"How lousy was the medicine then!" wrote Chekhov, after reading Tolstoy's description of Prince Andrei's death in *War and Peace*. It is terrible that these very words still apply to Russian medicine today.

Until the fifties, Russia was far ahead of the West in research of the autonomous nervous system. We understood the core differences between the autonomous system and the somatic one, and even in the absence of equipment, this led to testing on animals and, most importantly, progress in creating a systemic view of the autonomous system. In the last fifty years, a great deal of theses and publications on the subject came out in the Soviet Union, but they were mostly compilations and contained nothing new.

Once, a doctor did one or two X-ray studies of lymph drainage and then copied the rest from basic textbooks in order to write his dissertation. Then he used friends for opponents, bribed the officials, got his degree, and has been playing a professor his whole life, forgetting what he actually was. This was the M.O. for most Soviet "scholars."

This so-called professor and I had known each other since school. He could not forgive my success either in Moscow or in the West. His burning envy turned to hate as he tried to take revenge on me and discredit me. He spread a rumor that, instead of emigrating to Holland, I had been sent there by the KGB. This created great problems for me. At school reunions, I was given unpleasant looks, even from the alumni I had counted as friends. I was seen as a person who could not be trusted. And the "professor" reinforced that lie with little hints and winks.

Old Russian research is little known in the West. Only in the last twenty to thirty years have Western scientists become interested in the autonomous nervous system. A lot of data has been accumulated. Thanks to modern equipment and perfect technique, research is done in a single cell, even on the molecular level, and the electrical activity of each nerve fiber is recorded. Many refined experiments have been conducted that would remain a dream for Russia for many years or decades. But few attempts are made to analyze this data systemically.

Longevity and Health

I read in an American article on statistics of discoveries stating that, for every hundred researchers in the United States there are only two or three creative ones. I think that the ratio is the same in Europe. In my view, this is the reason for the absence of achievement in treating the autonomous nervous system: the absence of a systemic approach. Western researchers have to realize that only a systemic approach integrates the results of all analytical methods; otherwise, we will wait a long time for a breakthrough in treating many diseases. Systemic approaches are generally anathema in the States, where inventions are more highly regarded than theories. But this is good for business, and for ensuring that nitpicky studies of excruciatingly small quantities are what pass for research, thus ensuring another round of grant money. A systemic approach might put an end to that whole carousel. Thus, a systemic approach is not economically feasible in the Western economic system.

In Russia, an opposite situation existed. Advancement was political, not based on skill. Thus, one's fraudulent PhD was immaterial. It was merely a formality in getting some position in a ministry or on a committee, and thus securing semi-tolerable housing and a decent lifestyle. Here, triviality pays; there, corruption. In both cases, greed and the desire to be comfortable drives things. Both situations also pretend that a meritocracy is in effect, but this is not the case. This is why there is so little creativity in the United States or Russia at this point

If we adopt a ten-level system to evaluate the state of a science, then mathematics, physics, et al., will be on a very high level, say eight to ten. Yet Western medicine, both theory and practice, is no higher than levels five to six.

As for Russian medicine, it does not reach beyond level two. The Russian physicians who emigrated to the United States have learned a few things and thus stepped up to level three. This is why I persistently advise my friends and relatives who live in the United States against turning to Russian doctors. Of course, there are exceptions, but they are rare, which makes it hard to establish the level of skill of a given doctor.

I deem it necessary to summarize the reasons for my negative opinion of the level of medical science and practice in Russia. The low professional level of most specialists in science and technology in Russia is well-known, and medicine cannot form an exception.

Notes on Medicine

I have already written that medicine cannot be regarded as high science. Until the 1950s, Russian medicine kept up with world standards. Yet Stalin and his criminal Communist system destroyed the intellectual power of Soviet people—including prominent doctors—even in the period after his death. I have already written that in the last fifty to sixty years of Soviet/Russian history, not one major medical discovery has been made. Science was managed by Communists, who had a faint idea of medicine. The main factors in their pseudo-scientific research were falsehoods and lies. In fact, the whole nation, physicians included, was brought up in the spirit of total deception, and this holds true to this day.

The use of crib sheets was widespread in all the examinations in medical schools, and other colleges as well. This means that students—future physicians—were poorly trained and never mastered the medical sciences.

All major city hospitals have specialized neurological departments, headed by people who have to know their specialty, if only because this is where future neurologists come to do internships. For many years, I worked as a senior researcher at the Neurology Department of the Central Institute for Continued Medical Education at Botkin Hospital. We were not concerned with training, but rather with improving the skills of the heads of local neurological departments, where general specialists trained in neurology. I was shocked by these department heads' ignorance in neurology.

Our department head was Professor Chetverikov. Frankly, he was not the brightest neurologist, but against the background of his ignorant assistants, he seemed competent. When the administration failed to confirm me as the lab director, after Professor Chetverikov resignation in protest on my behalf, Petelin was who they chose to replace him. He was an absolute nonentity, ignorant in basic neurology. This was the man in charge of one of Russia's leading departments!

Here is another typical example of how doctorates were manufactured in Russia. Petelin compiled his thesis in front of me, using two hundred medical histories. The reader will find it hard to believe, but all two hundred were fabricated with wrong diagnoses. In order to keep this from his opponents, he changed the dates and the numbers for all two hundred histories This was how you became a Soviet professor!

I went through dozens of theses in either the capacity of opponent or reviewer. Almost all of them were falsified to a degree, sometimes so overtly that I had to withdraw. Yet all of them were successfully defended, and their authors joined the ranks of Soviet science. Falsifying the thesis was particularly widespread in Moscow and Leningrad, but the rest of the country also did it.

Once, I was reading lectures on neurology in Odessa. A local doctor asked me to review his doctorate. I was dumbstruck as I read it. It was typical schizophrenic writing, yet he already had two good reviews from major neurologists. I had to withdraw. A year later, he defended successfully and became a Candidate of Medical Sciences (PhD).

I do not cite these cases in order to discredit, insult, or wreak revenge. I am merely trying to illustrate the level of Russian medical science. A list of the "successful" thesis that I know about would make a separate book. Many doctors, hundreds or even thousands, emigrated, mostly to the United States, where they somehow passed the boards and now practice medicine. Even after further studies, their professional level remains very low. In effect, they were in business, sometimes quite successfully, and it is a reflection of our system of medicine that such a situation can persist

On the basis of many years of practicing medicine and conducting research in the former Soviet Union, and knowledge of work by many physicians who emigrated to the West, I can conclude that these doctors fall into several categories. Most of them are poorly trained, do not know much about practical medicine, and do not perform research. A few are trained well enough to match the international level of the seventies and eighties. As for their research, these "scientists" copy foreign publications and, for the most part, invent their own data.

There was a professor who once worked as a researcher at the Academy of Sciences Neurology Institute, the largest in the Soviet Union. He came to ask me to be his opponent at his doctoral defense. His subject was treating pathology of the carotid arteries with ionization, or a faint electric current. I was struck that he used a current that was tens of times higher than is permissible. I told him about it, he promised to check, and a couple of days later, he brought me a dissertation in which the electric current numbers were ten times lower than before. This is Russian scientific research for you. I refused to be his opponent. Nonetheless, he defended his dissertation successfully, became a professor, and enjoyed a high reputation among his colleagues. He even became an academician!

In the last ten to fifteen years, all medical institutes in Russia call themselves "academies," so scientific doctors automatically become professors, and professors become "academics," and all of this happens for money, bribes, and so on. Therefore, don't pay any attention to their titles.

There are two types of scientific people. The first type is highly educated and has a lot of knowledge but cannot produce many innovations. The second is educated well enough and has good doctors capable of producing very important medical ideas. Usually this second type has a difficult life, because other doctors are very jealous of them and make their life difficult. It is well known that the X-ray was discovered by a professor, Professor Rentgen, and he received a Nobel Prize for it. But his colleagues tried to undermine him, which caused him to become depressed. The end of his life was very bad. He was penniless, chastised by his colleagues, and discredited in the medical world.

In some of the areas where the Russian government's money was invested, most inventions were in the area of weapons development, including nuclear weapons. But even in this area the technology was not Russian but obtained by spies, who stole the technology from America and Europe. This is my opinion after sixty years of living in Russia. In conclusion, the knowledge of normal Russian doctors is at about the same level as a medic.

Years passed, and much water flowed down the Moscow River. My wife and I became Dutch citizens. The years I spent working for Terghuys and his TLB-Electronic were the hardest. I don't know what happened between him and Keis Kaismonbegen, but they became deadly enemies. According to Terghuys, Keis tried to blackmail him, because the former continued to pay his salary six months after Keis left the company, probably to keep him silent. I later heard that Keis had left for England, and I haven't seen him since.

Kaismonbegen disappeared, but Dr. Maylor reappeared. It turned out that Terghuys had had separate negotiations with him. Without my knowledge, they signed a partnership agreement, and Terghuys even gave him one of the EAG devices for research. After this, they suddenly asked to meet me, in order to find out my opinion on the best way to use the device.

I was outraged and told Maylor that I was not going to help. However, Terghuys managed to persuade me that being partners with Maylor was vital to the mass production and sales of the device. Finally, I gave a couple of recommendations. After all, I was curious about the results obtained by Maylor and his coworkers after six months of testing. When I asked him, Maylor abruptly declared that he would invite me to his laboratory only if he had need for me, and right now he didn't.

Later on, Maylor did need me, and invited me to the university. There, he set up a special research group that, parallel to the device, recorded the blood pressure, pulse, and other parameters. This was done by a highly competent woman who wanted to use the data obtained through the autonomograph for her own thesis. In theory, she could get interesting results, but what I saw did not reveal information on skin potentials. Everything was noisy and static, without one quality recording. This made Maylor lose his haughty air. He was confused and crushed, though he was to blame for creating this situation in the first place.

Still, I was interested in continuing the experiment. It appeared that the device needed improvements and adjustments. It worked fine in the absence of other equipment in the room, but when it was taken elsewhere, it recorded the surrounding noise. The group couldn't fix the problems, and they soon drifted apart.

Even before Terghuys gave Maylor the device, we built a four-channel sample that looked better, produced color charts, and worked, when it was in the separate room. When a businessman offered to buy the instrument, I warned him of its flaws, because I did not want to take the rap for Terghuys' errors. Of course, the desire to buy vanished. For various reasons, such as a lack of time from these companies our connections to Hewlett-Packard, Siemens, and others broke off as well.

During sixty years of experience as a neurologist, I was in the biggest hospitals in Moscow, Holland, Belgium, Switzerland, the United States, and other countries. I participated in many world congresses in many countries. I have to be honest; I came to the conclusion that treating brain diseases is very difficult, mostly near impossible. I developed a new complex of exercises with the head that improves the blood circulation in the brain. It is known that if the blood circulation is better, any damage can be treated more efficiently. These exercises don't require any additional medication. I will briefly describe only my idea of treatment, and later I will give an example of an old person who was treated, who without my help would have been untreatable.

Background Information

It is known that blood circulation is very complicated. I will briefly describe how the arteries supply the brain with blood. Though I will describe only part of the process, it will be clear that it is very complicated.

The nervous system is one of the most metabolically active tissues of the body, having an energy requirement ten times greater, proportionately, than the rest of the body. The adult nervous system requires about 18–20 percent of the total body's resting energy supply of oxygen, even though it represents only about 2 percent of the total body mass. Because it also stores very little glycogen but uses approximately 25 percent of the glucose consumed by the body, it depends on an adequate and constant supply of oxygen and glucose provided by the blood flow. Interruption of this flow to the brain for less than a few seconds results in loss of consciousness.

The brain receives its blood supply from two arterial systems: the carotid arteries and the vertebral-basilar arterial system. The two arterial systems are interconnected by anastomoses.

The carotid system consists of three major arteries: the common carotid, internal carotid, and external carotid. The common carotids ascend in the neck, where each divides into external and internal branches.

The two vertebral arteries join to form the basilar artery. The vertebral-basilar system of arteries supplies the spinal cord in part, brain stem, cerebellum, and some other important parts of the brain. This system has various branches.

The cerebral arterial cycle Willis is located at the basis of the brain. It interconnects the internal carotid arterial and vertebral-basilar arterial systems from which the anterior, middle, and posterior arteries arise.

The most common cause of narrowing of the lumen of the arteries that supply the brain is atheroma. This disease may affect the main arteries supplying the brain in their course through the neck as well as their course within the skull. Athermoatous degeneration of the cerebral arteries occurs most commonly in middle or old age. When actual blockage of an artery occurs, the effect will depend on the size and location of the vessel. The nerve cells and their fibers will degenerate in the avascular area. In patients with generalized narrowing of the cerebral arteries without blockage of a single artery, the brain will undergo a diffuse atrophy.

Cerebral veins are divided into an external or superficial group and a deep or internal group that both drain into the dural venous sinuses and ultimately into the internal jugular veins.

When organs have poor blood circulation, then they don't function well. Surgeons can inject blood or special liquids to increase the circulation. This technique cannot, however, be applied easily to the brain.

I have developed a noninvasive method to improve blood circulation in the brain. The method can be used many times during one's lifetime, several times a day, even. The method is very simple and can be used without medical help, although the patient must be instructed how to perform the exercises correctly.

People who are middle- or old-aged (more than fifty or sixty years) can suffer from arteriosclerosis. This can damage some of the brain structures. Even if we improve the blood circulation, we cannot revive atrophied brain structures. We can, however, improve the function of any remaining brain structures, which can compensate for the damaged ones.

I have developed special exercises for different people to train their blood supply to the brain. As with physical exercise in the gym, certain amounts of repetitions are not suitable for everyone.

As I have already written, neither medication nor equipment is necessary for these exercises because it is a biologically method which I called; "Biologically Pump Strengthening Intrabrain Blood Supply" The first few times a patient starts using this method should be under supervision and guidance from a doctor. After this, the patient can continue to exercise independently for the rest of his or her life. The results of these exercises will be quick, after several days.

Elderly people suffer from vertigo, deafness, tinnitus, intention tremor, dysarthria, dysdiadochokinesis (impaired rapid alternating movements), ataxia of limbs and gait, as well as other symptoms. These symptoms can be due to bad vascularization of the brain, infections, toxins, alcohol intoxication, trauma, and so forth.

Part of the exercises that I propose is connected to the carotid system. In these cases certain vessels may suffer from arteriosclerosis and therefore not function optimally. The exercises I described may however help the functioning of the vessels as they will be compensating for the vessels that are dysfunctional which results in an insufficient blood supply.

Another exercise can help different symptoms that arise from a damaged cerebrum, with its different structures. These structures may be damaged by a number of diseases: immune diseases, endocrine disorders, infectious diseases, tumors, trauma, or toxic agents, among others. The exercise that trains the carotid arteries can help strengthen these structures again. The method improves the blood circulation in the carotid, which will then stimulate the recovery of the damaged part of the brain. Besides stimulating recovery of the brain this also has a positive influence on internal organs such as the heart, stomach, lungs, etc.

My method can be used by healthy and young people as well, to prevent various illnesses, not only in the brain, but all organs. This is because all organs are regulated by the nervous system. When the nervous system is optimized, the regulation of the rest of the body is improved as well. Special exercises that I propose have the scope of improving the vertebral-basilar system. I understand that the vertebral-basilar system is only a part of the brain's blood supply, but it also has many connections with the carotid system. In elderly people, some vessels may not function optimally (such as in the case of arteriosclerosis); therefore, the connections between blood vessels may be insufficient. In such cases, these exercises must performed carefully, because the structures cannot supply sufficient blood.

During my career in medical practice I have seen many people with head trauma and brain damage. It is very difficult for this group of patients to receive the right treatment from a neurologist. I think that my new exercise method can solve a part of this problem. In the following paragraphs I will give two positive examples of my experiences with this treatment using my exercises.

A part of the exercises I propose are mostly connected to the carotid system. Certain vessels may suffer from arteriosclerosis and therefore not function optimally. In these cases, the exercises I describe may obstruct the functioning vessels, as they are compensating for the dysfunctional ones, resulting in an in sufficient blood supply.

There was a case of an eighty-year-old man. For the past twenty years, he had suffered from vertigo when standing up and moving from a horizontal position to vertical. As time passed the symptoms got worse, and during the past four months even turning from left to right while lying down created the dizzy feeling. When bending down to pick something up, or even turning left or right the man also suffered. He was unbalanced as he moved, forcing him to take much more time to walk. Some days he even needed a walking stick for stability. After just four or five days of performing the exercises I recommend, the man found that all his symptoms disappeared

An old men suddenly lost his consciousness, fell with his head on a stone floor which damaged his skull and caused bleeding in the brain. When he was brought to the hospital, the neurologists investigated him. They found a bleeding in the left temporal lob of the brain, several small bleedings in the frontal parts of the brain and some bleedings in other places of the brain. Next to this he suffered from a big skull fracture. After 3 to 4 weeks he recovered a little bit and I started to treat him with my new exercise because a shared friend asked me to help him. After 2 months of me treating him daily he recovered almost completely. The neurologist from the hospital could not believe that such a patient recovered after the severe damage of the brain he had suffered. I am convinced that there will not be many neurologists that can claim similar results using different methods.

The method can be used by athletes. This method can be used by millions of ill and healthy people alike. This will be an important step forward to a healthy long life, thus to Longehealthyvity.

At this moment I can't explain in detail how the exercises should be done, because I have applied for a new patent, and it has not yet been approved. When the patent is approved, a scientific article will be published, and I will be free to explain everything.

Chapter 14

How Kasparov Became the World Chess Champion (Unknown Facts)

In the early eighties, my wife and I happened to spend New Year's Eve at a wonderful Kremlin sanatorium in Sochi. A mere mortal did not have a chance of going there in the summer, but in the winter a few rooms were allotted to the Academy of Sciences, and that was how we ended up in this little haven.

Visiting there was the chairman of the Azerbaijan Republic, Gennadi Rsayev. Earlier, we had met him and become friends. Rsayev and his family came to visit us in Moscow, and we visited them in Baku and Zagulba, a favorite vacation spot for Soviet Azerbaijan rulers. One time Rsayev told me about an amazing chess player named Garry Kasparov, who became the junior world champion at seventeen and a grandmaster and national champion at eighteen. He was confident that Kasparov would become the world champion. He was right, and we did not have to wait that long. At the time it appeared unbelievable, though we constantly heard of this young chess player's amazing victories.

Rsayev introduced me to composer Leonid Vainshtein, Kasparov's uncle, and his wife, who were also vacationing in Sochi at the time. We spent New Year's Eve with them and have stayed friends ever since.

Time passed, and in 1984, a famous match started in the Columns Hall between Garry Kasparov and the world champion Anatoli Karpov. Of course, we were rooting for Kasparov. The details of this amazing and dramatic match can be found in Kasparov's book *Unlimited Challenge*, but a few things remain unknown to the public and chess lovers.

Everyone remembers that the beginning of the match turned out hopeless for the challenger—he was losing as he never had before. One loss, and then another, and a third. This was a shock! Many allowed that

Kasparov could lose the match, but not in a shutout. No one could understand what was happening, but Kasparov fans smelled a rat.

When the score was 5-0 in favor of Karpov, Leonid Vainshtein called me and asked if he could come to see me. He told me directly, "Something strange is happening to my nephew. He cannot sleep, eat, or prepare for games. Garry is in a state of severe depression and completely demoralized."

Clara, Garry's mother, confirmed this and stated her opinion that it was not worth continuing in this condition. It would be better to surrender and begin training for the next match.

I had followed Garry's progress for years and held a different view. If he lost this duel for the chess crown, he would not have this chance ever again. Of course, I shared this only with Vainshtein. Clara was already worried sick. She looked at me with hope and kept saying, "Tell me, doctor, can you help? Can you do anything?"

Before answering, I interviewed her in detail about the match, the place, the schedule, the conditions, the meals, the rest, the drinks, the personnel. The rivals stayed at the Russia Hotel, ate at the restaurant or used room service, but the food was always cooked by the hotel chef.

I discovered that, at the time of the match, Kasparov had been given a drink, and I wondered if something had been added to it. This would not seem improbable to anyone who knows the methods used by the KGB and was fully aware of the atmosphere of hate that surrounded Kasparov. I prescribed Garry some medication to boost his psychological condition and told Clara to cook his food, fix him a thermos with his own drink, and turn down everything else. To this day I am sure I was right.

On weekends my wife and I usually stayed at Aksakovo, outside Moscow, at a resort that belonged to the Foreign Trade Ministry. The director of the resort was Ignatov, a former deputy minister of the interior in Latvia, appointed to the job after his retirement. He was my patient and also offered me hospitality at his resort, whenever I came to see him. He sometimes asked me to take a look at someone.

Once, he asked me to take a look at the KGB official in charge of Aksakovo—that picturesque place had six to eight resorts. When I walked into Ignatov's office, I saw the KGB man chatting with a newspaper reporter from a Moscow paper. They were discussing the upcoming match between Karpov and Kasparov.

How Kasparov Became the World Chess Champion (Unknown Facts)

"Mind you, Kasparov's real last name is Vainshtein, and he is not Armenian, but a real Jew," the reporter said in a servile manner.

"Well, we won't let a Jew become a champion." Thus spoke the representative of the most powerful Soviet institution.

I heard this with my own ears. Hence, when the score was 5-0, I did not doubt that Kasparov had been drugged with some intellectual depressive medicine by State Security.

A few days after I had given the advice, Clara called me.

"Doctor, you worked a miracle!" she said happily. "Garry is revived. He got up at eight in the morning, rested, was cheerful, had a big breakfast, and sat down to practice chess. He is in a really good mood—not a trace of depression!"

Every day, Garry gained more strength and became more confident. He would visit our house with his friend, the actress Neelova, and we would watch television and talk. It was really interesting to talk to him about everything: literature, history, and philosophy. Sometimes I would also give him advice.

Karpov was already inviting friends and reporters to Game 31, where he planned to finish Kasparov. He arrived at the game wearing a new suit, ready to be decorated with a laurel wreath. Yet Garry managed to break away and fought to a draw. And then he won Game 32—his first win in this match. Karpov resembled a boxer who had just received a big punch in the head.

This was a psychological breakthrough that foretold the rest of the match. Of course, my advice helped Garry's condition, but perhaps the most important thing was that he stopped using the drink brought to him by the match organizers. His mood and energy levels improved radically in a mere two to three days.

After his victory, we met with Garry. He looked totally unlike his former depressed self. He was happy, sociable, and witty, and he unleashed an avalanche of ideas and information on us. He won another game and then another. Now a squabble started involving FIDE President Campomanes, the Sports Committee, and the Chess Federation. Now it was Karpov who was depressed and grew weaker day to day.

Out of the blue I got a call from my old friend Professor Gregory Kassil, who was then in charge of a laboratory of athletic medicine at a

191

research institute of athletic and biological studies. During the conversation he suddenly asked if one should take Nootropil for mental fatigue. I grew wary.

"What about Cerebrolizin?" he asked. "Is it effective in such cases?"

When Kassil mentioned Moriamin-M, I could not believe my ears. These were the same ingredients that I had prescribed for Kasparov. This was no coincidence, and there was only one way to explain it: Kasparov's hotel phone was tapped, my advice was "overheard," and now I was being asked if the same stuff that had helped Kasparov could also work for Karpov, who by then was in a state of physical and psychological exhaustion.

At the score 5-3 in favor of Karpov, the match was stopped. This outrage was committed with the full approval of the media, the public, the chess federation, and the Politburo. Along with other Kasparov fans, I was indignant.

It was hard to believe that a year later, when the two players met again, Karpov demanded that the new match should begin with the 2-0 score in his favor. This amazing psychological detail says a great deal about the twelfth champion of the world.

This time, Kasparov and his team trained more seriously. Our friend Gennadi Rsaaev led the team. When Garry and Rsaaev arrived in Moscow, they came to see me immediately, to discuss certain issues. Now they stayed at the Central Union Hall on Lenin's Avenue. The match took place in Tchaikovsky Concert Hall. The score difference was minimal. The pressure was relentless, like in a movie thriller, when you don't know the ending to the last scene. Kasparov won the last game and became the thirteenth champion of the world. Dozens of people rushed to congratulate him, and the fans carried his mother outside. A little later, Klara came up to me and said, "Eduard, no one knows how much you have done for me. If it were not for you, Garry would not have been a champion!"

On that day I heard many warm words from Leonid Vainshtein, Gennagi Rsaaev, and other close friends of Kasparov's. His book *Unlimited Challenge* provides many interesting details about the match. It is very odd, however, that Kasparov writes of a psychologist from Baku who helped him greatly, but not a word about me. I assume that the reason is that at the time, he and his mother were loath to identify with the Jews. Klara, always, and especially during the match, sought to position him as an Armenian and isolate him from Jewish relatives and friends, in the belief that his Jewishness could harm him. That is why Garry became Kasparov and not Vainshtein.

In his book, Kasparov returns to the details of his duel with Karpov and writes that many were against him and wanted to destroy him, but the times had changed. He was also absolutely opposed to the fact that we owed these different times to Mikhail Gorbachev. During one of his visits to Amsterdam, we talked politics, and he made his usual anti-Gorbachev remarks. I objected.

"Wasn't it Gorbachev who changed the world by starting Perestroika?" I said.

"No, it was Reagan's doing, not Gorbachev's," he said.

Of course, Reagan's role should not be belittled—it was his Star Wars program that forced the Soviet Union to step up its arms race with a weak economic base, but it cannot be denied that Gorbachev started Perestroika.

In 1991, Kasparov came to Amsterdam for a chess tournament. We had dinner but did not dwell on what happened in 1984/1985. Three years later, he came back to Amsterdam and visited me. I recorded his autonomography, and it confirmed Garry's unusual psycho-physiological makeup, and that was also very surprising for me. I will not go into details. I will only say that I had never seen anything like it.

Over many years we saw each other several times, and not once did he mentioned my assistance. Perhaps he did not like it that I had seen him in such poor condition at the time of the match. I do not need his gratitude, but it makes me happy to know that I had a part in his success. Without false modesty, I affirm that had it not been for my treatment and recommendations, Garry would not have become Kasparov, and the world would not have such a great chess master and the greatest champion in the history of chess.

Me with Isabella on our wedding day in 1952.

With Isabella, celebrating my seventy-fifth birthday.

My son, Leonid Lerner. professor of ophthalmology.

How Kasparov Became the World Chess Champion (Unknown Facts)

My daughter, Dr. Inna De Koning.

My dear sister, Dr. Maye Lerner.

My brother, Dr. Aron Lerner.

My birthday celebration. Here, I have reached the ripe old age of seventy-nine.

Chapter 15

Is Dr. Ernst August Klaus Graf Dangerous in Business?

In chapter 10 I wrote that I had found a partner and investor by the name of Dr. Klaus Graf. After thirteen years I still don't understand who Dr. Graf really is. Some of the information that I will give in this chapter about my ex partner Dr Klaus Graf I already gave in earlier chapters but I think it is important for the reader to understand who Dr Graf is and my relationship with him.

In 1997, we did not sign a contract because he wanted this relationship to be in his favor. He offered me an amount of money if I agreed to sign the contract. I then signed the contract that he had drawn up. Although ashamed, I signed because I needed Dr. Graf's financial support to finish my life's work. During the following years, I did everything for his best interests because he had invested funds into our research. During this time, I tried to save him money wherever possible. He was aware that I had saved him a great deal of money at that time.

This picture shows Dr. Claus Graf (R) and myself (L) during our positive period of our cooperation in 1998.

Longevity and Health

In the year 2000, after a successful period of ETDDS experiments, Dr. Graf changed his behavior towards me and this made me suspicious about his honesty. Between 2000 and 2006 he sent me many letters written in English asking me to sign these letters as soon as possible. His assistant, Mrs. Weiss, during this period, visited me in Amsterdam many times and she asked me to sign a number of letters immediately. She stated that she had to return to Frankfurt within the hour and that the letters were not of significant importance to me but if I did not sign the letters immediately Dr. Graf would stop sending me further funds and our companies in the Netherlands and Curacao would be lost. I felt I was being put under pressure by Mrs. Weiss to sign when she knew that I could not understand and read English. With some of the letter, Mrs. Weiss asked me not to put any dates on them. (i.e.: a copy of one of these letters without date as an example). At that time I trusted Dr. Graf, and so I did not make copies of many letters that I signed.

Is Dr. Ernst August Klaus Graf Dangerous in Business?

The undersigned	:	Dr. E.N. Lerner
Residing at	:	Arent Janszoon Ernststraat 17
		1083 GP AMSTERDAM
		The Netherlands
being the owner of	:	-three thousand nine hundred- (-3.900-) shares B and
		-one- share A m

PROGNOMED N.V.   ("the Company")

a Company established at Curaçao, Netherlands Antilles, herewith appoints

Mr. G. de Vries and/ or

Mrs. A. Steward

as his true and lawful attorney-in-fact to represent above mentioned shares with full power and authority, and with all rights and obligations connected therewith, at the extraordinary general meeting of Shareholders of the Company, to be held at Curaçao on   , in which meeting the items of the agenda as mentioned below will be dealt with:

- report of the management
- adoption of the financial statements and discharge of the management of the Company for its conduct of affairs for the periods ended December 31, 2002
- appropriation of earnings
- resignation and discharge as a managing director.
- transaction of any and all business, which may properly come before said meeting.

Any other proxy previously given by the undersigned is hereby revoked.

Date:

Place:

Signature(s): _____

    The above situation happened frequently between 2000 until 2006 without me understanding properly the meaning of the contents of these letters. Guarantees can be given by a number of people confirming that my English is bad, especially after the trauma of my head it deteriorated.

Bronwyn Tremalio
Schinkelkade 40/2
1075 VJ Amsterdam

26 September 2008

To Whom It May Concern:

This letter to confirm that I have worked for Dr. Eduard Lerner over the last ten years as a secretary/editor.

The main part of my job is to correct his written English into understandable text. Dr. Lerner speaks English poorly and he has great difficulty in writing and understanding complicated written English.

I am a native English speaker, which is why I am suitable to perform these tasks for him. Without my help he wouldn't have been able to write or fully understand any letters in English over the last ten years.

Yours truly,

Bronwyn Tremalio

To whom it may concern

I, E van Zanten, as a former employee of LMT herewith confirm that the English of Dr. E. Lerner is very poor and largely limited to understanding medical scientific papers and general conversation. However, business matters and juridical papers are for him very difficult to understand.

E. van Zanten, M. Sc.

Signed and dated the 19th of July 2003

Is Dr. Ernst August Klaus Graf Dangerous in Business?

 NEDERLANDSE ORDE VAN UITVINDERS

Dr. E. Lerner
A.J. Ernststraat 17
1083 GP AMSTERDAM

Postbus 10153
3505 AC Utrecht

Nijverheidsweg 16
3534 AM Utrecht

Tel: 030 246 7707
Fax: 030 244 0461

Email   info@novu.nl
Internet   www.novu.nl

ING Bank 69.32.35.195
Postbank 23.69.601

Handelsregister 40439741

Weesp, October 1st 2008

Re: Knowledge of English

Dear Dr. Lerner,

Recently you asked us to declare that your English reading and writing is such that complicated documents can't be interpreted well by you. We are always willing, knowing you for many years, to support you with this matter as a member of our organization.

Kind regards,

The letters that I had signed gave all the companies – Prognomed and Intrabrain – in Curacao, to a manager who could control the companies in any way he and Dr. Graf wished. Now I know that Dr. Graf has his ways to *persuade* anyone. This manager or another person from Curacao would fulfill all his requests because everybody trusted him or benefited from him.

During the occasions in which I was forced to sign many documents, tactics were used to put some pressure on me to sign quickly. In reality I was not given the time to discuss the contents of what I was signing with my lawyer. With the threat that my funding would be cut, I was coerced into signing documents that I didn't understand. If I wanted to continue my life's work, by pressing on with the research, I could only do what Dr. Graf, or Mrs. Weiss instructed as all the research was property of the companies. Without Dr. Graf's consent no action could be taken regarding any property of the companies meaning that I could not continue this research elsewhere, leaving me trapped.

Later, I asked Dr. Graf on several occasions to send me a letter with all documents attached, that he had asked me to sign over the years. I just wanted to understand what I had signed in the past. Graf however didn't show me anything. I still think there is a possibility that in one letter I signed, I gave up all rights for all 3 companies in Curacao and the Netherlands and that they might have been trying to sell the companies without my consent and my signature. I proposed to Dr. Graf that he could have 65% of all the companies and I would have 35%. Formerly, I had 65% and he had 35%. I further suggested that I would give him 80% or 90% of the companies and I would have 10% or 20% if he would send me copies of all my signed letters. This he did not agree to. This means that Dr. Graf could undertake some action with our companies, and that he didn't need my participation.

Graf asked me to find a buyer or investor for both projects, ETDDS and EAG. For the EAG project alone, he was looking to sell for €20 million that is stated in the fax shown below.

Is Dr. Ernst August Klaus Graf Dangerous in Business?

TELEFAX

TO   Dr. E. Lerner
            Lerner Medical Technology BV, Amsterdam

Fax-N°   003120 - 44 21 399

FROM   Dr. Graf

DATE   October 22, 2001

INCLUDING THE COVER PAGE YOU SHALL RECEIVE   1   PAGE (S)
IN CASE OF IMPROPER RECEIPT PLEASE INFORM:

Re: EAG
        Potential new investor

Dear Dr. Lerner,

my investment up to now in EAG (Prognomed SA) amounts to NLG 4.300.000,-.

In my opinion, your negotiations with the potential investor should be based on an actual market value of EAG of at least NLG 20 million.

Please keep me informed of the result of your forthcoming discussions.

Kind regards

(Dr. Graf)

I eventually found a potential new investor and Dr. Graf asked me to give him the name and contact details of this potential new investor. I think that Dr. Graf had asked me for these details in order to contact this investor and make some sort of deal with this investor without involving me.

In 2000 when all the experiments were successfully completed, I found a potential investors group. They promised to invest millions of Euros in our project. They were very interested in the project and together, the Director, manager and I went to Frankfurt to meet Dr. Graf. Whilst in the meeting Dr Graf asked me to leave the meeting so he could "freely" continue the meeting without me. I did not understand why he had requested me to leave the meeting but returned after one hour to hear that the potential investors would give me an answer the following day in Amsterdam. The next day they gave me a negative answer. They would not be investing any funds even though they had been very interested the previous day. I was very shocked to hear this as I did not understand what could have happened for them to change their minds. I now think that Dr Graf had, in some way during the meeting, whilst I was not present, some conversations with these new potential investors and I think he did this in order to continue business with them and without my involvement.

Dr. Graf had already done the same thing with some members of NOVU. When we had a meeting in Utrecht, The Netherlands with the Director and members of Novu, Dr. Graf stated to these people, in my presence, that I was a difficult person to work and discuss with. He said that I had called him many times, even at 11 p.m., to talk to him about something. This is not true because during the previous 10 years, he never gave me his telephone number. The only way I could get into contact with him was via his secretary during the day. After leaving a message, he only returned my call some several days or weeks later.

In 2006, without consulting me, Dr. Graf invited his lawyer from Boston, U.S.A, Mr. Shoolman, to prepare a manuscript about our projects. In the manuscript, Dr. Graf described our story differently to what was, in fact, the true reality.

Maybe I'm wrong but I think that Dr. Graf has in some way managed to sell the companies or patents or he owns 100% of our project for which I have worked all my life, forty years in Moscow and 19 years in Amsterdam. If he has sold the companies or the patents, then he has made many millions of Euros. Before, he asked 20 million NLG only for EAG which is far less profitable than DDS. In 2003 Dr. Graf estimated the value of our projects (ETDDS and EAG) amounting to 50 million Euros. The letter on the following page can confirm this.

Is Dr. Ernst August Klaus Graf Dangerous in Business?

T E L E F A X

To: Lerner Medical Technology Ltd. (B.V.)
     Attention: Dr. Lerner

Fax N°: 003120 44 21 399

From: Dr. Klaus Graf

Date: October 17, 2003

INCLUDING THIS PAGE WE TRANSMIT..... **1**......PAGE(S). IN CASE OF ERROR PLEASE CONTACT US IMMEDIATELY:

Dear Dr. Lerner,

In your negotiations with potential investors please consider the following:

My cash investment in the two projects – Electroautonomograph and Intra brain Drug Delivery through the Nasal Cavity is 3.4 million Euros. This figure does not include the cost of my staff (Mrs. Weiss, Mr. Weber, Mrs. Suraweera and Mrs. Tischhauser) during the last 7 years which I estimate at a cost value being € 1.5 million. You also have to take into account your accrued compensation of € 3 million as of December 31st 2003.

The total cash investment amounts to nearly 8 million.

Considering the exceptional progress we have made during the last 6 years, i.e. patents, positive experiment results and presentations in various international symposiums in Europe and the U.S. the value of the Lerner Group actually should not be below € 50 million.

Sincerely yours

(signature)

(Dr. Graf)

For this reason, we had a serious discussion and later I had to be nice to him otherwise he would have stopped sending me any more money. Until now, whenever he sends me money monthly, he now stated that this money would become a private loan to me, which is not true. I have an office, a translator, advisers and application for patents for which I also have to pay. These payments take up the largest part of this money. Besides this Graf asked me to pay an additional 5% interest over the money he had invested.

Longevity and Health

In 2002, I was so stressed because of this situation that I had a car accident. I contacted my insurance company to pay my damages. They asked me to send them a letter from my company confirming I was unfit to work. I requested that Dr. Graf would send a letter stating that I was not fit to work as a co director. He sent me a letter stating that because of the accident he was terminating my position as Director of the companies (see below copy of letter). Although, he knew I was quite capable of carrying out my functions. Dr. Graf also terminated his position as a Director of our companies in Curacao, handing the position over to a company named Fides N.V.

After the letters of Dr. Graf I was interested in my real position in our companies so I asked Fides N.V in Curacao to inform me about the status. I have received a letter from them recently that states clearly that: *"Please note that you have never been appointed as one of he managing directors of the above companies. Since incorporation until December 31, 2003, N.V Fides and Dr. K Graf were managing directors of both Companies"*. This for me was really strange because I was always told that Dr. Graf and I were always the co directors in both companies.

I hope that the reader can see that this is an example of Dr. Graf not being honest to me.

Is Dr. Ernst August Klaus Graf Dangerous in Business?

DR. KLAUS GRAF

RUOSTELSTRASSE 15
CH-8835 FEUSISBERG/SCHWEIZ
TEL. 0041 1 787 68 90
FAX 0041 1 787 68 91
E-MAIL : buerodrgraf@active.ch

To whom it may concern

In July 1997, the undersigned and Dr. Eduard Lerner formed Lerner Medical Technology BV, Intrabrain NV and Prognomed NV in which Dr. Lerner holds a 65% and the undersigned a 35% interest.
Subject of the companies was to improve, register and market various inventions of Dr. Lerner.

To achieve these results the undersigned financed the activities of the companies until 2002 additionally with loans in an amount of € 3.200.000.--

Until 2002 the companies improved there objections substantially and were close to sign Joint ventures agreements with International pharmaceutical companies.

After an accident of Dr. Lerner which happened in May 2002, Dr. Lerner was no longer able to perform his activities as Managing Director of the Group.
As a result, the progress of the reached results stopped and could not be further developed as it was foreseen and the companies did not reach income.
In the actual situation the company is not in the position to repay my loans and the company most probably will loose there high income projection.

Feusisberg, 23. October 2004
Dr. Klaus Graf

Not long ago, I sent a letter to Dr Graf to try to persuade him to continue our work again. Although our partnership did not end well, I still needed him to finish my project. The letter below shows this.

From: E. Lerner, February 4, 2008

Dear Dr. Graf,

Holland is a small country and everybody has their own opinion, which will be very dangerous for us. I see only one solution. Until this Friday – 8 February, before I will have an appointment with the Director of the NOVU, I need a letter from you that I have some shares in both companies. I do not mind if you have 65% and I have 35% and that I am a co-director as I was before. Otherwise the Director of NOVU will not contact you or find new investors. Also, please send a letter to the Managing Director in Curacao, instructing him to change immediately all paperwork to state that I am a co-director. Ask him to send me a letter stating the changes. Any new potential investor will check the company's history worldwide. If this is not done, they will not trust us and I will not be able to persuade them otherwise. Besides this, they will insist that you give a guarantee that I will never be asked to sign letters to Curacao stating that I give all my rights to the manager in Curacao. It should be stated that he can only do the taxes and other official documents in Curacao. Over the last five to seven years, I have done virtually nothing concerning these projects because I did not have any money. Now, with new investors, we have the possibility of completing the projects in the next two or three years, resulting in you receiving many billions of dollars.

No doctor in the world will treat patients with Alzheimer's or other diseases without me as the author of the method. Firstly, I have to do research testing, a publication, get a FDA license and then I can start to treat, together with other doctors, patients. You are a very clever man and if you do not immediately do what new investors ask, we will lose all of the projects.

If, for one or two years, we do nothing, other doctors will write new patents and they will do everything without us. Billions of dollars will go to them but not to you.

One advisor told me that I can cancel all of my patents and with my knowledge, write new patents but I would never consider such a thing because it would be against everything that you and I do.

Is Dr. Ernst August Klaus Graf Dangerous in Business?

Dear Dr. Graf, at the end of our lives, let us finish these projects as soon as possible, in order to treat millions of patients and to get billions of dollars. I am sending this letter because I trust you very much.

Sincerely

Edward Lerner

Dr. Graf, resigned his position as Managing Director of the company Intrabrain in Curacao. This meant that Dr. Graf and I were no longer managing directors of these companies. All the rights we owned, Graf gave, as I have written before, to people in the companies in Curacao.

Intrabrain International N.V.
Pietermaai 15
Curaçao, Netherlands Antilles

Mr E.A.K. Graf
Ruostelstrasse 15
CH-8835 FEUSISBERG
Switzerland

Ref: Resignation Intrabrain International N.V. (the " Company")

I, Mr. E./.K. Graf, residing in Switzerland, hereby resign as Managing Director of Intrabrain International N.V. as per December 31, 2003.

Place:
Date:—
By: _____

Mr. E.A.K. Graf

All decisions were given to the managers in Curaçao and they have all the power, they are now in a position to make any decisions they wish to make. I think Dr. Graf is still able to influence the decision making with the Curaçao managers.

Reflecting back on why exactly Dr. Graf resigned, I realized that if he distanced himself from the companies officially, he could also not be held accountable for any decisions they make regarding sales of shares, patents or any other movement within the business.

Is Dr. Ernst August Klaus Graf Dangerous in Business?

I became tremendously nervous when I found out that he was taking away both our projects DDS and EAG, for which I had worked on from the beginning in Russia and for a period of 19 years in the Netherlands.

In 2006, I was so nervous that I suddenly lost consciousness; I fell and hit my head. I was taken by ambulance to the Academisch Medisch Centrum (AMC) Amsterdam. The doctors discovered that I had cracked my skull and had caused some damage to a part of the brain. Day by day I recovered and as a neurologist I treated myself. After one and a half to two months I recovered completely and all the doctors were stunned at my quick recovery.

I also show here a letter from my house doctor, Dr. R. van Coevorden regarding my present health situation and the reason why I had been suddenly become extremely ill.

Longevity and Health

SCEN arts
Consulent Palliatieve Zorg

Amsterdam, Tuesday, 21 October 2008

L.s.,

For the past 16 years I am the family doctor of prof. dr. Eduard N. Lerner, d.o.b. March 5th, 1929.
18 years ago he arrived from the Sovjet Union. At the time he was the head of the biggest neuroscientific laboratory in Moscow. Upon his arrival in the Netherlands he started to conduct original research in the field of neuroscience and farmacology. 12 years ago he started to cooperate with a business partner from Frankfurt, Germany. The first 5-6 years this partner invested a lot of money into the research with very good results. They established 3 companies. The last 5-6 years the relationship with this partner deteriorated. Because of this prof. Lerner suffered from nervousness, had sleeping problems and developed high blood pressure.

In august 2006 prof. Lerner suddenly fell unconscious, fractured his skull and suffered from cerebral hemorrhages. He was hospitalized for 2 weeks in the Academic Medical Center hospital. He remained with memory problems and physically with problems of sleeping, and mood changes c.q. depression.

Presently prof. Lerner is under a tremendous amount of stress because, as he puts it, his German partner

Willem Coumou
Manueel therapeut

Pieter Nortier
Manueel therapeut

Marsha Pinedo
Kindertherapeut

Marja Schraal
Sociaal Psychiatrisch
Verpleegkundige

A.J.Ernststraat 173, 1083 GT Amsterdam the Netherlands
Tel.nr. +31-(0)20-6441627; Fax.nr. +31-(0)20-6421201
Postgiro 4605860 Bank ABN-AMRO 493057870

tricked him into signing contracts (in English, what he does not fully understand anymore, lost ability to read, write and properly speak), taking away his shares in the laboratories or most of it. At least that is his impression and are his worries.

Prof. Lerner explicitly asked me to write this down, as he is not able to do this himself also because I as his physician have witnessed all of the above.

R.S. van Coevorden, MD

In 2008, Mrs. Weiss became very rude and unpleasant. She would never reply to any of my questions. Whatever I asked her, she was very critical and became quite annoyed. Maybe she felt she could be disrespectful towards me because she knew of Dr. Graf's negative opinion towards me.

Usually, Dr. Graf sent me 6,000 Euro per month. During the first several years, he sent this money to the company account. Later, he then sent 4000 Euro to the company account and 2000 Euros to my private account. During the last several years, he has sent 2,000 Euro to the company account and 4000 Euros to my private account. His secretary on sending the money stated in a letter that the money in question was a loan to the company and to me. Dr. Graf charged me (as I have already stated) 5% interest on this loan. I know Dr. Graf, and I am afraid that maybe he is quite capable of asking me to return this investment (loan?) at any time.

Dr. Graf was always looking for anything negative or bad in the way I functioned in my job. In November 2009, his lawyer, Mr. Shoolman, from Boston, whom Dr. Graf invited to Frankfurt, called me and asked me some questions. Firstly, he asked why I got a patent to deliver drugs not through the nasal cavity but through the spinal cord several months ago. I told him, the application for this patent, I sent several years ago to Washington. At that time, I paid for the patent and I have now received it. Now I understand that this way of delivering this drug to the brain does not improve our project. Therefore, I only put my name on the patent.

Mr. A Crawford, from the Washington patent office rang me and told me that Dr. Graf's lawyer, Mr. Shoolman and Mrs. Weiss had rung him and asked him if I had made any mistakes with the patent and if I had paid in time etc. etc. They asked him many questions in order to try and find fault in my actions. Mr. Crawford told them that any questions they have, they had to refer to me as I was the author of the patents. Of course, some patents were not published in time; some of them were lost because Dr. Graf had not paid for them when I sent him a bill from the patent lawyer.

Previously, I wrote that I would explain more about how Dr. Graf can win people for his own gain. For example, the administration and financial advisors in Loenen aan de Vecht, Netherlands, Hulsman-Roestenburg, are our financial accountants. We made an agreement with the above company that they would work for us for 3% of our profits. Dr. Graf, without telling me, suddenly paid them 10,000 Euro. Mr. Roestenburg was very happy and he stopped talking to me. I could not even ask him to do any work for me. This made it very difficult for me to have any business relationship with Mr. Roestenburg.

Usually Dr. Graf tries to distance himself from bad situations and he asks his assistant Mrs. Weiss to handle any business that he does not want to get involved with. At one time, we had an appointment. Dr. Graf, Mrs. Weiss, Mr. Weber and myself. Suddenly, Dr. Graf said, "we have to give something to Dr. Lerner because he needs something to eat". I was in shock because I own 65% of the companies and Dr. Graf wanted to give me only "something to eat". It seems that he thinks he has complete control of the companies.

Of course, as an author, I can cancel all my patents for EAG and DDS as I have mentioned already. I have worked for them all my life and it is therefore not easy choice for me to make. If I decide to cancel these patents, I know this would have an extreme adverse financial effect on Dr. Graf. Although, I would like to cancel these patents but I

Is Dr. Ernst August Klaus Graf Dangerous in Business?

know I cannot do so as I feel that these patents are so very important for millions of patients in the future who cannot be cured without my methods. I am presently working on new research and in the near future, if I need to take out a patent or patents they will belong solely to me, then I might consider cancelling my previous patents which belong to our previous companies.

The following patent applications were not granted due to lack of funding from my partner Dr. Graf.

CA2238055 A1 19970529
AU7693296 A 19970611 [AU9676932]
EP0869829 A1 19981014
DE69216404 T2 19970724
US5482052 A 19960109 (appl. 08/140,058)

As I had said before in the book, several years ago I had a contact with Mrs. Ponomareva, the assistant of Mr. Abramovich (Moscow). I sent her many documents about my inventions (e.g. Business plans, etc). Dr. Graf was aware of this and promised to make a deal with Mr. Abramovich. Until now, I have never seen the results of this and Dr. Graf told me smiling, that there had been no deal. I still don't trust this answer.

Dr. Graf has nearly a hundred companies and many hundreds of employees, so I can imagine, in my experience, how many people have questions from doing business or working with Dr. Graf. Leaving aside the many problems between him and his employees, it is important to note that the larger problem at hand was Dr. Graf's lack of interest in furthering medicine. Stronger still, his interest was focused on making money through medicine, not on helping treat patients. It took a few years for me to open my eyes to this however, by which point I was willing to give him up to 90% of the profit from the work I had done. I was willing to do so because I was interested in finishing my life's work and of course receiving money for it, although not the primary goal. I haven't worked at all for the past 10 or 11 years, because Graf didn't want to finance the research. To oppose such a rich person as Dr. Graf was impossible for someone with my limited resources. The only option I see around the situation is to find a new investor to finance the last few years of research. Only then will

217

we be able to start applying the technology to treatments of patients worldwide.

At the end of 2009 I received a telephone call from Dr. Graf, and he was excited. "I have you and the business plan in my hands, and thus I will pay you €6,000 monthly till the end of your life" he promised. After 3 months he stopped sending this money. On the 19th of November 2010 I was in Frankfurt and he told me that he couldn't remember his promise. This, I have learned, is typical of Dr. Graf, his words offer everything, but in his actions he falls short. It could be that he didn't do this because he wanted to cheat me or because his memory is not good anymore.

In 2010 My lawyer, made an agreement between me and Dr. Graf stating that I could buy all shares, including all patents etc., which belong to, Intrabrain N.V. and Prognomed for one Euro. In the beginning Dr. Graf agreed with this, but for several months he didn't sign this agreement.

When I was with him in Frankfurt in November 2010 he had prepared a completely new agreement without my lawyer. Dr. Graf assured me that his agreement was nearly the same as the first one. He also said that if I didn't sign his agreement he would withdraw all financial support to me. Dr. Graf knew that my English is very bad but he still gave it to me without my being able to read it.

After having signed the agreement I asked him how much he would pay me per month. He told me that our cooperation was finished and wouldn't pay me anything. After a long discussion he told me he would pay me 2,000 Euro's a month for the first three years. In reality I have only received one payment and nothing else. This is a typical example of Dr. Graf. He never does what he promises.

I understood that soon I will have to stop my research which I have spent 60 years working on. Near to 40 years of which were spent in the Soviet Union and around 20 of which were spent in countries in the West.

Due to my experience I can compare the different ways of working between the "gang - communists" country called the USSR, and the honest and decent West.

In Moscow I was the head of one of the biggest laboratories in the world, and one of the youngest doctors of medical science, which in the Soviet Union was the equivalent of a professor. I have obtained

Is Dr. Ernst August Klaus Graf Dangerous in Business?

excellent results from my work. Due to the envy, enmity and hatred of other doctors, even my friends, my laboratory in Moscow was destroyed (as shown in chapter 5). I was also fired as head of the laboratory. Luckily with the help of the government (for whom I was a doctor) I moved from the medical field to that of research at the Academy of Science of the USSR. Later I went to the West, hoping to continue my research. Even though I was over 60 years old, the Dutch doctors gave me a job. I have previously described my career in the West. After having met Dr. Klaus Graf, he became my partner and investor. He didn't destroy my laboratory as happened in Moscow, but with his hidden actions I think he achieved the same as the USSR gang. I think that due to his actions I suffered from a head trauma, but I managed to treat myself successfully. Now I am healthy.

Any company who bought programs from the companies from Curacao (Intrabrain and Prognomed), couldn't do much because all the knowledge necessary to successfully apply the technology is still in my head and it will take an additional 2 years together with me to finish this project and after that millions of patients can be treated. Without me, the project will and cannot be finished within 15 - 20 years and the buyer can lose all of his investments that can sum up to more than 50 Million Euro according to Dr Graf.

After working with Dr Graf for 13 years, I still have no answers on some questions:

Did Dr Graf cheat me?

Is Dr. Graf dangerous in business?

Did many millions of people die and did many more become ill, because Dr. Claus Graf stopped the development of my method to treat these people?

Maybe some readers will be able to answer my questions.

Epilogue

The first time I came to Holland, I was introduced to an English doctor. While we sat in a café in Amsterdam, I told him of two of my inventions. He listened to me attentively and said, "Dr. Lerner, you are past sixty. These ideas are so colossal that you will never have the energy to actually implement them. It is practically impossible for a person coming to the West without connections or knowing the language. You will risk your own health and maybe life, but the con men you will run into will cheat you anyway. I am a disinterested party, and I am just telling you as a friend: give it to someone local and let them do it. You will be proud that these ideas are yours. Do not get involved—you can't win."

I heard him out, but deep down inside I could not agree. Surrender was not my way. Now, without false modesty, I can draw some tentative conclusions. I am proud that I have developed, constructed, and tested my device, the electroautonomograph, in the biggest clinics in Europe and America. It is the only tool in the world that has the ability to study the condition of the autonomous nervous system directly and noninvasively. It is the only tool to diagnose patterns that have been scientifically linked to various states of diseases.

The electroautonomography method allows the selecting of candidates with healthy autonomous nervous systems for such fields of endeavor as space travel, aviation, and naval work, as well as work in countries with hot and cold climates. It can be used in sports medicine for selecting prospects. This device alone can diagnose the potential, from an early age, for developing hypertonic disease, gastro-intestinal diseases, and others. It can also be used in the pharmaceutical industry for drug testing and in veterinary medicine for diagnostics.

The electroautonomograph has a 4–5 percent margin of error, which makes it more precise than the polygraph, where the margin is 20–25 percent. The new type of autonomous system activity diagnosis device discovered and patented with the electroautonomograph will allow for the creation of a noninvasive device for regulating the blood flow, bringing it up when it goes down and vice versa.

Longevity and Health

My second and most important project, as I have already written, is the delivery of drugs into the brain through the nasal cavity by using iontophoresis. During these last few decades, scientists have been trying to overcome the blood-brain barrier. Sometimes they have succeeded, but never so directly as via the nasal cavity. Illnesses like Alzheimer's, Parkinson's, epilepsy, strokes, brain tumors, psychiatric illnesses, and drug, alcohol, and smoking addiction (which affect one billion people in the world) are hard to treat, since medication introduced as pills or injections gets into the blood but rarely makes it to the brain. Since treating Alzheimer's (which affects more than four and a half million cases in the United States) costs over $100 billion dollars a year, and every fifth American suffers from some kind of damage to the nervous system, one can imagine the scale of potential use of this treatment and its economic effect.

It is important that the method is not invasive, painful, or expensive, and that patients can use it themselves without the help of doctors.

In the last ten years I have received over sixteen patents for these two projects from the United States, Europe, and other countries. I have given many presentations at international conferences in the United States and Europe.

The ETDDS research started at a large clinic at one of the oldest universities in Europe, in Belgium. There, we conducted very expensive experiments on the PET (positron emission tomography), a type of nuclear magnetic imaging device. We tested with this nuclear magnetic material because we could easily track it after the tests using this imaging device. The nuclear material showed up on the images. The doctors, Professor Dr. K. Van Laere, Professor Dr. L. Mortelmans and Professor Dr. G. Bormans, from the Leuven University Hospital, were so interested that in 2003, we started the ETDDS and PET experiments at ten o'clock at night and ended at two or three o'clock in the morning. They argued that we could not work during the day because the experiments were not paid for. After the first test, the radiologist Professor Dr. G. Borman was ecstatic. We had achieved the same results I had seen before myself when I performed these experiments in Russia. My colleagues showed me the picture on the computer screen and said it was magnificent. I was not surprised, as I had expected nothing less.

Epilogue

The following day, Professor van Laere told me that he had been wrong and the result was still unclear. After the second experiment, the story repeated itself; at first the result was interesting, and the next day I was told it was dubious again. After a series of five experiments I was promised I would get the results for the whole series; then I was told all the experiments showed negative results. When I asked about a controlled experiment, they muttered something incoherent. When I offered to pay for it out of my own pocket, I was turned down.

The entire situation appeared strangely familiar. When we began the research, a representative from the hospital's patent department came to see us. They showed a lot of interest and promised to draft a contract. After a few months, despite my reminders, the draft never materialized. An old story repeated itself: after familiarizing themselves with my work in detail, they decided that I was not needed anymore.

Because of my bad English I was not able to write an article all by myself so I asked colleagues to be co-authors. An example of this is an article that I have written together with Gregory R Stewart from the company Genzyme (US). The article was published in the Journal of Drug Targeting, June 2004, vol 12(5) page: 273-280. In the end he placed himself as the only corresponding author.

In 2007, I was introduced to Dr. Ir. Ward Mosmuller, MBA, director of technology, Transfer Office VU (Amsterdam, The Netherlands), who promised to find investors for my inventions. He invited me many times to discuss all the details of my inventions. Once, he told me that there would be a biotechnology congress in Rotterdam, attended by many investors. He asked me to go to Rotterdam, and he would help me find an investor. By this time I did not visit congresses anymore as a presenter but purely as a visitor. In Rotterdam he introduced me to a young girl and later to a very old man. I spoke with them for five minutes and realized that they didn't understand anything that I had explained to them and that Mr. Mosmuller was not honest because he didn't actually know any real investors. I think on the basis of my meetings with him and his incompetency at finding investors that his intention was only to understand the details of my invention.

In 2008 I called a Dutch journalist, Hans Knoop, with whom I had met twelve years before. He was very interested in my invention and promised to find an investor. In the meantime, Professor Booij had told me that Hans had been excluded from the Dutch Journalist Association for some professional mistakes he had made. After several days, Hans called me and told me that the honorary president of European Journalists, a German who was in charge of the Benelux, Mr. Helmut Hetzel, would come to visit me. When we met and I showed him my invention, he became extremely excited and told me that it was so important that its details must not be published in a newspaper. He promised to find a big company that would be interested, within several days. He told me that he had a good relationship with Philips and that in one or two days they would invite me to discuss my invention and that he was sure that Philips would cooperate with me. I waited for one or two weeks, but nobody called me. After one month I called Mr. Hetzel, and he told me that Philips could not give an answer immediately but they would think about it in the future. It has been almost a year and I have not heard from them.

I can't exclude that Mr. Hetzel didn't persuade the company Philips to take my invention.

I want to repeat again that some of the world's biggest universities, hospitals, and pharmaceutical companies have tried to steal my ideas and eliminate me from the projects. But it was not possible for me to fight them because they are very wealthy, they have the backing power of good lawyers, and they are always convinced that they are in the right. They always thought that they had the advantage of being strong and backed up by lawyers, while I was just a simple scientist from Russia.

I have had several requests from individuals to cooperate with them to further my research. After giving these people the information they required, they disappeared. One such man was Michael K. Vredegoor, a Dutch businessman who promised to put me in contact with a Canadian investor who was interested in my work. After having given him insight into my work and provided him with extensive results from the ETDDS and EAG inventions, he disappeared. Later, I found out that he had been involved in dubious practices in Holland and had no option left but to move abroad, and so he went to Greece.

Before this incident, however, I had an experience with an American who told me there was a professor in Copenhagen who would be able to help me with my research. After explaining my work to this professor, he informed me that the workload involved in this project was too much for him and he would therefore not be able to participate. After one year had passed, his assistant from Copenhagen University published an article about drug delivery through the nasal cavity using iontophoresis!

On December 17, 2008, there was a biotechnology congress in Utrecht, Holland. The previous month, there was a symposium in Leiden, Holland, for people who wanted to present their invention at the congress. Ten people, including myself, made a presentation. Among those, the only one out of the ten not accepted was my work. Later, I understood that the leader of the biotechnology symposium, and the leading manager of research and inventions at Leiden University, intended to keep my inventions for themselves. At the symposium, I gave them my memory stick, which included not only my presentation but a lot of confidential data, and they copied the entire thing. After a few minutes, the leading manager of R & I at Leiden handed me my coat and showed me the door.

There are medical professors at Leiden University who specialize in drug delivery through the nasal cavity, but without iontophoresis. With my information, they can use my invention. Besides that, I made the mistake of showing Alie Tigchelhoff my business plan about my second invention, electroautonomography. I asked her to sign a nondisclosure letter, which she refused to sign. She kept the business plan for one month and then sent it back after I had reminded her many times. She also sent me another nondisclosure letter, which was completely different. These two ladies took confidential information about my two projects, and perhaps one day, I will see whether they gave my inventions to someone to complete and to use in practice.

In this book I only present you some examples of colleagues trying to profit from my inventions but such situations happened very often.

One of the many doctors working against me was the Professor from the Amsterdam Medical Centre. He had an especially good relationship with Professor Mathius from England, Professor Hilz from Germany and other professors from other countries. He created and led a blockade against me in the medical field.

Longevity and Health

Twelve years ago, when my wife fell ill with the Guillain Barré syndrome, she was treated in the neurology department of the Academic Medical Centre in Amsterdam. After a number of mistakes made by some of the neurologists, my wife was near death. Before this, I had good relations with the neurologists at the AMC, but when I saw that their errors might result in my wife's death, I had a choice to make. I could either remain on good relations with these doctors and watch my wife die, or I could pull the alarm bell and save my wife at the cost of my professional relations. I sent a letter to the Ministry of Health describing the mistakes I had observed. This stirred up the necessary commotion, after which the doctors followed my instructions and my wife recuperated. Unfortunately, the doctors involved in this affair turned against me.

The Dutch Association of Inventors (NOVU) made an agreement with the University of Rotterdam that they would prepare a business plan for me. It became clear from this plan, which is partially explained in chapter 11 and 12, that if my method of drug delivery were introduced, even for the treatment of a mere eight diseases, and only in the United States, it would not only treat millions of patients more effectively, but would also yield many billions of dollars a year for the investors. If this method will be implemented all over the world for a wider scope of diseases, the result would be stunning.

Despite the hardships that have hindered my progress, I continue to work and hope that I will succeed in putting my ideas into practice. I am already past eighty. My work enjoys demand in the West, and the world of neurology has acknowledged my work. As you have seen in chapter 13, I'm currently working on a new invention for special exercises to treat certain brain illnesses.

As explained in chapter 15, during the last 12 years I had a partner Dr. Claus Graf. He was my investor and in the first years of our cooperation he invested money and I had very good progress with experiments and my research for ETDDS and EAG. But soon after, Dr. Graf stopped to finance and I could not finish all my experiments and start to treat patients. Next to that he made it very difficult for me to look for new investors.

In the end of 2010 Dr. Graf prepared an agreement between us to finish our cooperation. With my bad English I couldn't read the agreement and after some pressure from him, I signed the agreement.

Epilogue

In the book I presented a letter from Dr. Graf in which he estimated the value of the projects to be around 50 (fifty) million euro. If Dr. Graf sold my inventions of which I still have 16 patents, I'm interested to know, how many millions he got for my inventions?

I can imagine that the Billionaire Dr. Graf saved a lot of money by not investing more in the inventions and therefore millions of patients died and many hundreds of millions became very ill.

I am not trying to present myself as an ideal person. In life I was forced to do a few indecent things in defense against the attacks. I don't believe in conceding to evil, so if I get slapped on the cheek, I don't turn the other one.

I will admit that I am not very modest. During my entire career I have been very successful which made my colleagues and even my friends jealous.

Going back to academic Pavlov's words about "nonstop thinking"—it follows me constantly, like a shadow. His words meant that a scientist must not stop thinking. And I do not, no matter if I am eating, or talking with someone, or watching television, or walking outside. Every morning I take a walk for four or five kilometers along the canal, in the woods, enjoying the splashing of water, the murmur of the leaves, and the chirping of the birds. In those moments I think up ideas very easily. Inspiration doesn't strike only poets, artists, and composers; a few times it has visited me as well. This is the absolute truth, and it helps me live, work, and enjoy life despite the difficulties.

Most countries will be interested in my methods—all insurance companies and many doctors as well. If I manage to bring them to fruition, then the near future will bring about a real revolution in various fields of medicine, not only neurology. At the same time, I realize that publishing this book will not make my life and work easier.

At the end of my life, I am compelled to agree with Professor Joachim Volted, whom I quoted earlier:

"Scholars who have achieved success do not have a pleasant life; they are envied, feared, but most often, they are hated. They have known envy and hostility from their colleagues, and even from their friends."

Appendix

Dr. E. Lerner's International Patents and Publications

A Part of the Patents:

Device for enhanced delivery of biologically active substances in an organism
WO9718855 A1 19970529
US2002068080 A1 20020606 (appl. 09/197,133) (US6410046)
US2002123678 A1 20020905 (appl. 09/077,123) (US6678553)
US3191426 A1 20031009 (appl. 10/393,254) (US7200432)

Method and device for enhanced delivery of a biologically active agent through the spinal spaces into the central nervous system of a mammal
US2002082583 A1 20020627 (appl. 10/050,183) (US6913763)
US2002183683 A1 20021205 (appl. 10/051,817) (US7033598)

Apparatus particularly for use in the determination of the condition of the vegetative part of the nervous system
WO9219172 A1 19921112
EP0585303 B1 19970102
US5522386 A 19960604 (appl. 08/140,056)

Method of determining the sensory functions
WO9219154 A1 19921112

Administering pharmaceuticals to the mammalian central nervous system
US6410046 B1 19981120 (09/197,133)

Device for enhanced delivery of biologically active substances and compounds in an organism
MX240861 19980520 (984007)
US6678553 19980520 (09/077,123)

Method and apparatus for enhanced and controlled delivery of a biologically active agent into the central nervous system of a mammal
US7033598 B2 20020118 (10/051,817)

Apparatus and methods for measuring autonomic nervous system function
US6490480 B1 20000914 (09/661,353)

Publications:
Lerner EN, van Zanten EH, Stewart GR.
Enhanced delivery of octreotide to the brain via transnasal iontophoretic administration.
J Drug Target. 2004 Jun.12(5):273–80.

  - Many Russian publications before 1990. These are not cited here.

Reports:

Title: Electroautonomography: A Diagnostic Tool for CFS
Date: November 9–11, 1995
Congress: First Congress on Chronic Fatigue Syndrome and Related Disorders

Title: The Electroautonomograph: A Tool for a Diagnosis and Interpretation of CFS
Date: October 13–14, 1996
Congress: American Association for Chronic Fatigue Syndrome Research Conference

Title: Intracerebral Drug Delivery
Date: December 1–4, 1997
Congress: Fourth Congress of the European Society for Clinical Neuropharmacology

Title: Electroautonomograph (EAG)
Date June 1–4, 1997
Congress: Tenth European Congress of Physical Medicine and Rehabilitation

Title Electroautonomography: A Non-Invasive Apparatus to Asses Autonomic Nervous System Function
Date: October 9, 1998

Appendix

Congress: Clinical Autonomic Research Society (CARS) Annual Meeting—Inaugural Meeting European Federation of Autonomic Societies

Title: Electroautonomography
Date: October 30–November 1998
Congress: IX International Symposium on the Autonomic Nervous System

Title: Existence of Periods of Highly Synchronized Electric Activities of Different Organs during Sleep
Date: July 1-3, 1999
Congress: The First European Federation of Autonomic Societies (EFAS) Meeting

Journals:

Clinical Autonomic Research

Volume 8 Number 5 October 1998
Title: Electroautonomography
Page: 310

Volume 9 Number 1 February 1999
Title: Electroautonomography: A Non-Invasive Apparatus to Assess Autonomic Nervous System Function
Page: 42

Volume 9 Number 4 August 1999
Title: Do We Know Everything about Skin Potentials?
Page: 222

Volume 10 Number 3 June 2000
Title: The Age of Independent Autonomic Prevalence Assessed by Electroautonomography (EAG)
Page: 162

Volume 10 Number 4 August 2000
Title: Fast Skin Potential Analysis: A Step toward Non-Invasive Microneurography

Longevity and Health

Page: 246

Volume 10 Number 4 August 2000
Title: Spontaneous Electrodermal Activity in Patients with Autonomic Dysfunction
Page: 262

Volume 9 Number 5 October 1999
Title: Existence of Periods of Highly Synchronised Electric Activities of Different Organs during Sleep
Page: 305

Volume 10 Number 6 December 2000
Title: Spontaneous and Evoked Skin Potentials in Patients with Autonomic Dysfunction
Page: 381

Volume 10 Number 6 December 2000
Title: Polysystemic Sympathetic Decrease after Local Ganglion Stellatum Blockade
Page 386

Volume 11 Number 3 June 2001
Title: Transnasal Iontophoresis, a New Non-Invasive Drug Delivery System for the Brain That Circumvents the Blood-Brain Barrier
Page: 205

Volume 11 Number 3 June 2001
Title: Is Chronic Fatigue Syndrome a Disorder of the Autonomic Nervous System?
Page: 205

Volume 11 Number 3 June 2001
Title: The Effects of Diabetes on the Central Autonomic Nervous System Assessed by Electroautonomography
Page: 205

Volume 12 Number 4 August 2002

Appendix

Title: Transnasal Iontophoresis, A Non-Invasive Brain Drug Delivery System That Bypasses the Blood-Brain Barrier
Page: 303

Volume 12 Number 4 August 2002
Title: The Next Step toward Non-Invasive Microneurography
Page: 328

Volume 13 Number 5 October 2003
Title: Epidural Iontophoresis—A Method of Delivering Drugs into the Central Nervous System
Page: 391

Volume 14 Number 5 October 2005
Title: Does the Electrocardiogram Have Biological Noise from the Stomach, Skin, and Other Organs?
Page: 319

After 2005 I could no longer participate in international symposia due to lack of funding for travel costs from my partner Dr. Graf.

Longevity and Health

 NEDERLANDSE ORDE VAN UITVINDERS

Postbus 5005
1380 GA Weesp

Van Houten Industriepark 35
1381 MZ Weesp

Tel. 0294 252 862
Fax 0294 252 865

E-mail info@novu.nl
Internet www.novu.nl

ING Bank 65.12.35.195
Postbank 25 65 601

Handelsregister 40349147

Weesp, August 29th, 2007

TO WHOM IT MAY CONCERN

Re: Inventions of Dr. E. Lerner

As The Dutch Association of Inventors we evaluate approximately 1.000 inventions every year. Sometimes we encounter truly remarkable things. In this framework we would like to introduce Dr. E. Lerner.

Dr. Lerner has worked and lived in Holland for the past 17 years. Before he lived his entire life in Moscow, where he was chairman of one of the world's biggest medical laboratories and head of the Neurological Department in a hospital.

During his membership of our organization he developed into a respected member. He received the Dutch Top Invention Award of our association, as well as several other awards, medals and letters of recommendation from European and American universities. He is the owner of 30 patents in a.o. Europe, The United States, Japan and other countries. In The Netherlands he has continued to develop two important projects: an apparatus to diagnose many illnesses of the nervous system and an apparatus to deliver drugs directly to the brain which treats many illnesses such as Alzheimer's disease, Parkinson's disease, drug and alcohol addictions and more than 50 other diseases.

Dr. Lerner performed many experiments in the best laboratories in Europe an achieved important results. The projects are 70% complete.
Several large companies such as Novartis, Johnson and Johnson and Genzyme, were prepared to finance his research, but Dr. Lerner preferred a more independent way of financing his final research phase, for which he has the next two to three years in mind.

In medical history it is well known that such big discoveries have a new epoch in medicine. At the beginning of last century a Dutch doctor - Dr. W. Einthoven - discovered and developed an EKG for which in 1924 he won the Nobel Prize. An English man - Mr. A. Fleming - discovered penicillin and an American - Dr. S. Waksman - discovered streptomycin, for which he also got a Nobel Prize.

Using Dr. Lerner's methods hundreds of millions of ill people can be cured. Herewith these **unique** inventions are so important for practical medicine that nomination for a Nobel Prize is quite well possible.

Knowing that an up-to-date business plan is of the utmost importance, we recently introduced Dr. Lerner to the Erasmus University in Rotterdam, where the NOVU provides guest lectures. A plan was made by two graduates and was reviewed by us on request of Dr. Lerner. Although the projected total profit of this remarkable invention is extremely high, it is our opinion that the basis of the figures is correct and that the assumptions are quite reasonable.

Please, feel free to contact me if you need any additional information.

Yours sincerely,

W.E. Pyzel
Managing director

NOVU
Dutch Association of Inventors

Member of IFIA

Appendix

Postbus 105
3632 ZT Loenen a/d Vecht

FAX   FAX   FAX   FAX

tel. +31 (0)294 232862
fax +31 (0)294 234304
e-mail: 101550,1713@compuserve.com

KvK Rotterdam - V 346121

Faxnummer   : 020 646 60 02
Bedrijf   :
T.a.v.   : Dr E. Lerner

Datum   : March 31, 1997
Betreft   : Translation press-release

Van   : W.E. Pijzel

Rough translation of the press-release done by February 24th, 1997. The award related to this title will be handed over at the NOVU Annual Meeting on the 21st of May in The Hague.

The Board of the NOVU has established by the beginning of 1997 the title "Dutch Top Innovation".

This title will be awarded to an innovation of one or more NOVU members with a invention of a level far over an average and of an interest that is far more than that of the inventor. Next to that the innovation has to be successfully tested by a prototype or clinical research. For a patent has to be applied. The board asks the meaning of an expert.

For the year 1997 the title is awarded to:

> Dr Eduard Lerner from Amsterdam with his invention
> 'Non-invasive Intracerebral Drug Delivery System"

Summarised this invention means that no longer medicaments have to be applied by injections or oral. They will be transported directly to the brain, without comming into the bloodstream. Side-effects belong to the past. Illnesses like Alzheimer, MS, Parkinson and epilepsy can be treated successfully. Herewith this invention belongs to one of the most interesting in the field of medicine of the last decennia and can be compared with the invention of antibiotics.

The expert on wich the choice of the NOVU is based is Prof. dr. L. Booij from the University of Nijmegen. The meaning of Prof. Booij is supplemented.

More details: NOVU
                Dr Lerner
                UvN

Mr. D.M. Gvishiani
State Comittee of Science and Technology
M o s c o w / USSR.

Moscow, Dec. 15, 1969

Dear Mr. Gvishiani :

      You are probably aware of the interest shown by various research institutes in the computers that can be provided by HEWLETT-PACKARD

      One case which has resulted in us receiving an order is that of Professor Lerner of the Institute of Psychiatry. We have been working with Dr Lerner and his engineers from the Institute od Automation and Telomechanics for the past six months and have now developped a very advanced system for use with a variety of patient monitoring equipment, a Nucleo Chicago Gamma Camera and a Micro Densitometer. The computer part of the system is now ready for delivery.

      . This advanced computensed faculty allows all the Physiological Parameters to be continously monitored as well as permitting the carrying out of advanced research projects.

      To our knowledge there are only perhaps three or four systems in existence at present. Specific location being the Mount Sinai - Hospital in Florida, USA, and the Hammersmith hospital in London, U.K.

      Whilst we are prepared to deliver and install a working system it would be virtually useless unless Dr. Lerner is provided with adequate software and hardware support from competent sources. This being the case we would appreciate efforts by your committee to ensure this.

      Yours sincerely,

      HEWLETT - PACKARD S.A.

This letter confirms the interest of Hewlett-Packard, as described in the text.

Appendix

Academisch Ziekenhuis Nijmegen

Anesthesiologie

Huispost: 520

Radboud centraal
Geert Grooteplein 10
Postbus 9101
6500 HB Nijmegen

Telefoon: 024-3614553
024-3614406
Fax: 024-3540462
024-3540524

Hoofd: Prof.dr. L.H.D.J. Booij

Polikliniek Pijnbehandeling:
Telefoon: 024-3614576

St Radboud

onze referentie   datum

November 6, 1996

betreft

To whom it may concern

During the last several months in the Pharmacokinetic Laboratory of the Institute of Anesthesiology we performed experiments to prove the invention of Dr E. Lerner of the *Non invasive intracerebral drug delivery*.

The obtained results flabbergasted me. If to take into account that he Blood-Brain Barrier is of the same nature for all animals, including humans, I am convinced that Dr Lerner indeed invented a non-invasive method to deliver drugs direct into the brain by bypassing the blood-brain-barrier.
The concentration of the delivered '*lead*' drug methylprednisolone in the brain is 10-50 (-100) times higher than after oral, parenteral delivery in the maximum oral dose of for instance 1 gram of methylprednisolone.
In the same time the concentration of methylprednisolone in the blood is very low or even absent.

This invention opens a new way for curing such difficult diseases like Multiple sclerosis, Parkinsonism, Alzheimer disease, Edemas swelling after brain surgery, strokes, trauma, Encephalitis, Meningitis, Epilepsia, etc.

This invention is really revolutionary, it will change the medical field and the pharmaceutical industry, Dr Lerner will receive exceptional credit for it.

Yours sincerely

Prof.Dr. L.H.D.J. Booij

Stafleden:
Dr. B.J. Arnold
Dr. H.B.M. van Beem
Dr. S.J.P. Crul
Drs. S.J. Danduan
Drs. D. van Dijen

Dr. R. Dirksen
Drs. R.T.M. van Dongen
Dr. J.J. Driessen
Drs. R. Eijk
Drs. B.G.F. Heeters
Dr. H.J.M. Gielen

Drs. C. Gors
Dv. P.H.H.M. de Groud
Dr. Kho Hing Gwan
Drs. J.B.M. Harbers
Dr. H.A.W.M. Hasenbos
Dr. J.G.C. Lerou

Dr. T.H. Liem
Drs. J.M.J. Mourisse
Drs. Ph.C.S.M. Poraisen
Dr. E.N. Robertson
Drs. D.G. Snijdelaar
Hw.drs. P.M. Vermeulen

Klinisch lysicus:
Dr. J. van Egmond
Klinisch farmaceut:
Dr. T.B. Vree
Hoofd beheerstaken:
Hw. Ir. D.H.A. Hubers

01-96

04/11 '96 14:34 FAX 0243540331   KLIN FARMACIE   ☑001

4-11-1996

Intracerebral delivery of drugs.
Methylprednisolone

E Lerner[1], T.B.Vree[2]

1) A.J. Ernststraat 17, 1083, GP Amsterdam,
2) Institute for Anaesthesiology, Academic Hospital Nijmegen, Sint Radboud, Geert Grooteplein Zuid 10, 6525 GA, Nijmegen, The Netherlands.

Dear Eduard,

Leo Booij was flabbergasted (very moved) by the results. This means a revolution in Neurology. Be very careful with Ciba, you can operate from a much stronger position. For methylprednisolone we can try to interest Upjohn or Merch Sharp & Dohme. When Michel Vredegoor is back from the states, please come to Nijmegen with him to discuss the possibilities with our Contractdepartment, or we must hire a professional contractor. Millions are involved.

Again a Russian revolution, 1917-1997?

Stay calm, cool

Tom

Dr. Tom Vree is a colleague of Professor Booij and worked with me on the experiments. This letter again illustrates that they were very much interested in my inventions or the financial consequences of my inventions.

Appendix

Fax

Ciba-Geigy AG, Basel, Schweiz
Ciba-Geigy SA, Bâle, Suisse
Ciba-Geigy Limited, Basel,
Switzerland
CH - 4002 Basel

AN / TO	VON / FROM	
Dr. Eduard Lerner	Name	Dr. P. van Hoogevest
A.J. Ernststraat 17	Ref.	PH 2.31 - Nr. PH_00285
NL 1083 GP Amsterdam	Standort Location	K-401.369
	Fax No.	+41 61 / 696 69 81
	Tel. No.	+41 61 / 696 56 51
	Datum Date	May 6, 1996
Fax No: 31-20-6488002	Seitenanzahl No. of pages:	1

Re: Invitation for a Seminar on "Non-Invasive Intracerebral Drug Delivery Systems"

Dear Dr Lerner

I refer to our phone-conversation of today, and would like to confirm our invitation to you, to come to Basle and give a seminar for a selected audience on your inventions on May 29th, 1996.

Your travel costs will be reimbursed by us and the seminar and further discussions will take place under secrecy agreement.

I would greatly appreciate if you could confirm your visit to us by fax and inform me about your exact time point of arrival in Basle and the title of your seminar as soon as possible.

In addition, I would greatly appreciate if you could inform me a little bit more about your inventions by means of available non-confidential information.

I am looking forward to hear from you soon.

With best regards.

Yours sincerely

Dr. P. van Hoogevest

CC Dr. G. Haas
   Dr. H. Rettig
   Dr. K Liechti

> The biggest proportion of my e-mails and conversations were conducted in English, so most of the time people helped me with writing the e-mails and reading what was written.

Longevity and Health

Division Pharma
Division Pharmaceutique
Pharmaceutisels Division

ciba

Fax

Ciba-Geigy AG, Basel, Schweiz
Ciba-Geigy SA, Bâle, Suisse
Ciba-Geigy Limited, Basel,
Switzerland
CH - 4002 Basel

AN / TO		VON / FROM	
Dr. E. Lerner		Name	Dr. P. van Hoogevest
A.J. Ernststraat 17		Ref.	PH 2.31 - Nr. PH_00334
NL 1083 GP Amsterdam		Standort Location	K-401.369
The Netherlands		Fax No.	+41 61 / 696 69 81
		Tel. No.	+41 61 / 696 56 51
		Datum Date	August 12, 1996
Fax No: 00031-20-646 6002		Seitenanzahl No. of pages:	1

DRAFT OF STATEMENT FOR YOUR EVALUATION

Dear Dr. Lerner

Please find enclosed a darft of the statement.

Re: Statement on the Importance of Intracerebral Iontophoresis (non- Invasive intracerebral drug delivery systems) for Drug Therapy

Dear Dr. Lerner

We have carefully evaluated your invention "non-invasive intracerebral drug delivery system" for which you received a prestige Award from the Town of Geneva and a Bronze Medal at the 24th International Exhibition of Inventions" in April 1996 in Geneva.

We agree with you, if successful this invention could have great consequences regarding treatment possibilities for diseases related to the central nervous system.

We would like to perform jointly with you a feasibility study in animals under the conditions described in the draft Agreement of July 1996. However it should be realized that further development is dependent on the outcome of this feasibility study. In addition, the development of this new drug delivery strategy and device demands a careful assessment of potentials and risks to obtain approval by regulatory authorities and will therefore require considerable time before final commercial realization.

Yours sincerely

Dr. P. van Hoogevest

Appendix

Ciba-Geigy Limited
CH-4002 Basel
Switzerland

Dr. E.N. Lerner
A.J. Emststraat 17
NL 1083 GP Amsterdam
Netherlands

Dr. P. van Hoogevest
PH 2.31 - Nr. PH_00354
K-401.369
Telephone(061) 696 56 51
Fax (061) 696 69 61

September 23, 1996

Re: Secrecy agreement

Dear Dr. Lerner

Herewith I would like to thank you for your visit to us last Friday, September 20th, 1996.

Although, we did not reach an agreement, at least we had to chance to create an atmosphere of mutual understanding and trust.

I hope, that you now realize that we will do our utmost to reach an agreement with you. In that respect, you will soon receive a draft study protocol and draft contract on the joint bodydistribution studies in The Netherlands with rabbits.

I enclose the secrecy agreement.

The participants of the technical discussion were:

Dr. Gerard Flesch, Biopharmaceutical Department (specialized in bioanalytics, PK and PD)
Dr. Helmut Schuetz, Biopharmaceutical Department (responsible for animal experiments)
Dr. Steffen Lang, Bioanalytics Department (responsible for iontophoresis)
Dr. Sietse Wouters, Bioanalytics Department (responsible for iontophoresis)

Regards

Yours sincerely

Dr. P. van Hoogevest

> **This document shows that this company was still interested in doing joint tests on rabbits, although I did not close a deal with this company at that time due to financial reasons.**

FAX

Ciba-Geigy Limited
CH-4002 Basel

To		From	
Name	Dr. Eduard Lerner	Name	Robin Walker
Fax No.	0031 20 6466002	Fax No.	++41 61 696 4720
Reference		Tel. No.	++41 61 696 6431
Company		Department	Corporate Unit Legal Services
Total pages	17	Date	October 18, 1996

Dear Dr. Lerner,

I am pleased to confirm that the financial terms for the experimental work to be conducted by the University of Nijmegen have been agreed at DFL 40,000. A draft Agreement is being despatched to Dr. Brooij this afternoon.

I should be grateful if you could confirm that the University does not have any rights to the invention as a result of the work they have done with you to date, and also that they too are bound by a Secrecy Agreement.

At a meeting attended by Dr. Haas, as well as Drs. Schuetz, Degen, Schächter, van Hoogevest and Csendes the nature of the proposed collaboration with you was discussed. I came away with the impression that all present were interested by the technology and looking forward to future cooperation with you.

I enclose the latest version of the draft Agreement. Although this looks rather a long document, it now contains the latest version of the Study Outline prepared by Dr. van Hoogevest and the usual lawyers' "boilerplate" at the end. You will notice that we have included the agreed payment for your work with the University. Please check the bank account details, which have been taken from your letter of the 26th September.

We are already giving some thought to the terms of the future Agreement. However, before we go too far down this road, we need to be clear as to the patent situation, and it would be extremely helpful if you could get a copy of your patent application to Dr. R. Schächter of our Patent Department as soon as possible. He will be here next week but away the week after.

I hope the draft is understandable to you. If you have any queries concerning the interpretation of any parts of it, please do not hesitate to get in touch with me. I should feel much happier if you were to seek the advice of a lawyer on the terms of the document.

We welcome the opportunity to work with you on this extremely interesting project. Hopefully you too are looking forward to a relationship with Ciba, one of the major CNS companies in the world.

Dr. van Hoogevest and Dr. Csendes join me in sending their best wishes,

Your sincerely,

Robin Walker
Division Counsel

cc. L. I. Csendes
Dr. G. Haas
Dr. R. Schächter
Dr. P. van Hoogevest

Vanderbilt University Medical Center

Italo Biaggioni, M.D.
Bonnie K. Black, R.N., N.P.
Randy D. Blakely, Ph.D.
Thomas L. Davis, M.D.
André Diedrich, M.D., Ph.D.
Emily Garland, Ph.D.
Nancy Keller, Ph.D.
Satish Raj, M.D.
David Robertson, M.D.
Rose Marie Robertson, M.D.
Paula Yamhure, R.N.

Autonomic Dysfunction Center
Phone: (615) 343-6499
Fax: (615) 343-8649

Nashville, 10th March 2006

To Whom It May Concern,

I have used Dr. Lerner's invention for the last six years. Therefore I can declare as follows:

Over several decades scientist and doctors tried to develop such a device like Dr. Lerner's invention. In the last two years, new devices appeared on the market measuring the autonomic regulation - but only of the heart. Similar devices to Dr. Lerner's invention, which can simultaneously register a wider spectrum of autonomic regulation (not only the heart) including gastrointestinal system, respiration, etc. do not exist currently. Therefore, the introduction of Dr. Lerner's device into medical practice will be a breakthrough in diagnostic pre-clinical illness. It will open new ways of prognosis for several diseases before the illness develops.

It is known that EKG was developed by a Dutch doctor Einthoven, for which, he received a Nobel prise in 1924. It is also known that EKG has a diagnostic value only when the heart is already damaged. It cannot make a prognosis for heart problems.

The EAG apparatus is able to make a prognosis of heart disease. The introduction of Dr. Lerner's device in medical practise can be as important as the introduction of the EKG.

The assessment of sympathetic and parasympathetic autonomic function by the EAG device could be useful to predict sudden death of adult people. It also will help to estimate the risk for hypertension in earlier stage before manifestation of diseases and therefore it can be used for improvement of preventive treatment. Another example of the importance of the EAG is the application of this device in patients with diabetes. Many

246

Appendix

millions of people suffer from diabetes. Diabetes often leads to a complication of gangrene and amputation of the limb. Doctors can only see this complication in the final stage of the disease when the damage is already done. The EAG can measure the dysfunction of peripheral nerves in the early stages of diabetes before gangrene appears. This can be used to make the treatment of diabetes effective for the preventing of amputation for hundreds of thousands of people. This will have a big social and economic effect.

The next steps in the development of the EAG device are improvement in hardware and software according to my recommendation. This will make the EAG very unique and it will be used by many medical professionals.

André Diedrich, M.D., Ph.D.
Res. Asst. Professor of Medicine and Biomedical Engineering

Longevity and Health

UNIVERSITAIRE
ZIEKENHUIZEN
LEUVEN

Nucleaire Geneeskunde

Prof.dr. L. Mortelmans, *Diensthoofd*
Prof.dr. P. Flamen
Prof.dr. S. Stroobants
Prof.dr. K. Van Laere

Telefoon
– Inschrijvingen: (016) 34 37 01
– Secretariaat: (016) 34 37 14/15
Fax: (016) 34 37 59

Leuven, 28/05/03

Research agreement between

Dr. E. Lerner, representative of Lerner Medical Technologies Inc., Amsterdam, here after called "Lerner Technologies"

and

Prof. Dr. K. Van Laere
Prof. Dr. G. Bormans
Prof. Dr. L. Mortelmans

Department of Nuclear Medicine, Leuven University Hospital, hereafter called "the host institution".

Concerning : study of transnasal facilitated cerebral transport using iontophoresis

Project proposal : This study aims at demonstrating presence or absence of a transnasal route to deliver drugs to the brain, in the pilot study using PET with a non-BBB-permeable radiotracer (F-18-benzoic acid) in a rabbit model. This study will be carried out at the host institution according to the proposal as agreed by the Ethical Commission for Animal Experiments

Responsibilities :
Organisation of and carrying out the study (purchase and management of rabbits, instrumentation for tracer delivery, rabbit preparation, PET scan and data analysis) will be the responsibility of the host institution.

Lerner Technologies will provide the iontophoresis instrumentation as stated in the ethical protocol, and will carry out the iontophoresis experiments.

Use of data :
The data from the study will be the joint property of both Lerner Technologies and the host institution and can be used for scientific publication by the host institution and by Lerner Technologies in case of positive or negative results.

Future collaboration:

In case the host institution decides to continue the study based on positive results, it will have the first choice of utilization of the iontophoresis instrumentation in joint imaging research experiments for other animal or human studies and tracers as it deems suited. The economical rights for future benefits are solely reserved to

U.Z. Gasthuisberg, Herestraat 49, 3000 LEUVEN

248

Appendix

Lerner Medical Technologies. Collaboration regarding imaging studies with oth r centers will only be obtained after mutual consent.

This agreement should be considered as a first working draft, and is to be superseded after revision by the Leuven Research and Development department regarding possible financial implicat ns concerning future collaborations.

Hereby all partners confirm the agreement of the items stated above,

Dr. E. Lerner

Date: 28-5-2003

Prof. Dr. K Van Laere

Date: 47/5/203

Prof. Dr. G. Bormans

Date: ....

Prof. Dr. 1. Mortelmans

Date: ....

E. Lerner

From: "Wise Young" <wisey@pipeline.com>
To: "E.Lerner" <e.lerner@wxs.nl>
Cc: "Giacin, Kenneth" <kgiacin@corus.jnj.com>; "Patricia Morton"
Sent: dinsdag 22 januari 2002 16:20
Subject: Re: experiments

Dr. Lerner,

Please don't apologize for your English. You were very clear.
Patricia and I very much enjoyed the meeting. I learned a great deal
and believe that you are one of the most creative scientists that I
have ever met. In my opinion, your ideas concerning transnasal
iontophoretic application of drugs and autonomic nervous system
monitoring have many important commercial applications. As you know,
Ken Giacin is a very good friend and I would very much like to help
Johnson & Johnson evaluate the technology. I am also mindful of the
confidential nature of our discussions.

As I indicated at our meeting, we are currently assessing the effects
of intrathecally and intraventricularly applied inosine to stimulate
corticospinal tract regeneration. We are collaborating with Larry
Benowitz and his group at Harvard on these experiments (see attached
abstracts). This would be a good opportunity to compare transnasal
iontophoretic and intraventricular administration of inosine. I have
been thinking not only of comparing the two routes of administration
but also experiments that would rigorously assess mucosal histology,
passage of fluorescent dyes of different molecular weights, and viral
passage into the brain associated with different levels of transnasal
iontophoretic currents. Some nasal mucosal damage may occur in the
presence of high drug concentrations, particularly at high levels of
current or voltage; if so, it would be important to determine the
safety limits of transnasal iontophoretic currents. Although I
personally think that viral passage into the brain is unlikely, a
convincing way to prove this is to apply a known virus that can
transfect brain cells and leave evidence in the form of fluorescent
gene expression in the cells.

Wise.

Appendix

Dr.E. Lerner

From: "Wise Young" <wisey@pipeline.com>
To: "E.Lerner" <e.lerner@wxs.nl>;
Cc: "Glacin, Kenneth" <kglacin@corus.jnj.com>
Sent: donderdag 23 mei 2002 17:31
Subject: Confidentiality Agreement

Dr. Eduard Lerner

Dear Eduard,

I apologize that it took me a while to read through the
confidentiality agreement that you sent. I am very sorry but it will
not be possible for me to sign the agreement. The agreement
contains terms requiring assignment of intellectual property rights
and restrictions on disclosure of our research. As a professor at
Rutgers University, I am not permitted to give away publication
rights or potential intellectual property rights of the University,
particularly stemming from research carried out at the University.

I would have no difficulty agreeing to "strictly protect all
confidential information and make absolutely no use or disclosure of
it, except if and as previously authorized in writing by LERNER".
However, I cannot agree to the following terms, as stated in the
agreement:

* As I interpret this statement, this means that we would have to
disclose any other ongoing research in my laboratory that be related
to the technology or interests announced by LERNER. Since I am bound
by other confidentiality agreements, this would not be acceptable.

Wise.

Wise Young PhD MD, Professor II & Director
W M Keck Center for Collaborative Neuroscience
Rutgers, State University of New Jersey
604 Allison Rd, Piscataway, NJ 08854-8082
tel: 732/445-2061, fax: 732/445-2063
email: wisey@pipeline.com, young@biology.rutgers.edu
web: http://carecure.rutgers.edu, http://sciwire.com

March 2, 2000

Lerner Medical Technology, Ltd.
A.J. Ernststraat 171
1083 GT Amsterdam
The Netherlands

Attention: Dr. Eduard Lerner

Re: ALZA Contract # 20000100

> Alza was one of the companies that was very interested in cooperating on my invention. I did not mention this in the book, but this letter shows the company's interest.

Ladies and Gentlemen:

This Letter Agreement (the "Agreement") will confirm recent discussions between ALZA Corporation ("ALZA") and Lerner Medical Technology, Ltd. ("LMT") regarding ALZA's interest in evaluating certain business and technical information of LMT relating solely to LMT's proprietary method and devices of iontophoretic and/or phonophoretic intranasal delivery of a therapeutic or other agent of interest to or through the brain (the "Technology") which LMT considers to be confidential and proprietary for the purpose of a possible collaboration. The definition of "Technology" shall specifically exclude (i) any information relating to devices or methods of electrically-assisted transdermal or transmucosal delivery of a therapeutic agent for local or systemic effect which can be used on a body surface other than the nasal mucosa, and (ii) any information relating to devices or methods of electrically-assisted delivery of a therapeutic agent through the nasal mucosa for local or systemic delivery to sites other than to or through the brain.

Very truly yours,

ALZA CORPORATION

Carol A. Gamble
Senior Vice President and Chief Corporate Counsel

APPROVED AND ACCEPTED THIS *10* DAY OF *March*, 2000.

Lerner Medical Technology, Ltd.

By: */E. LeRNeR/*

Title: *Director*

G:\MMD\ALT\SRR

ALZA CORPORATION   1900 CHARLESTON ROAD   P.O. BOX 7210   PHONE 650.564.5000
                       MOUNTAIN VIEW CA 94039-7210   http://www.alza.com

Appendix

ISTITUTO DI CLINICA NEUROLOGICA
DELL'UNIVERSITÀ DI BOLOGNA

LI 11.03.1999
VIA UGO FOSCOLO, 7 (Porta Saragozza)
TELEFONI 58.50.91 · 58.50.53 · 58.51.58
40123 BOLOGNA

Professor Lerner
Lerner Medical Technology Ltd (BV)
AJ Ernststraat 171
1083 GT Amsterdam
The Netherlands

Dear Professor Lerner,

On 25th February I was pleased to have the opportunity to evaluate in my laboratory a new device called electroautonomograph (EAG) which you designed to investigate the influence of the autonomic nervous system on different organs and on the skin potential.

We studied one subject with multiple system atrophy of the parkinson type (male, 56 years) associated with autonomic failure (MSA + AF), one subject with pure autonomic failure (PAF) (male, 57 years), one subject with essential hyperhydrosis in remission phase (male 50 years) and a female with epileptic seizures for which right frontal lobe surgery was performed (46 years).

I would like to indicate that at this stage I would be supportive of the possibilities of cooperation , following some of the aspects that we outlined and discussed.

There has been an increasing amount of interest in the ANS both from clinical and research point of view. Therefore any new approches such as EAG have considerable potential in terms of their commercial application and implication.

I look forward to continuing to be in communication.

Yours sincerely,

Pietro Cortelli MD, PhD

253

Dr.E, Lerner

From: "Mathias, Christopher J"
To: <e.lerner@wxs.nl>
Sent: dinsdag 29 oktober 2002 10:58
Subject: AAS meeting etc

Dear Professor Lerner

I was sorry for a variety of reasons not to have been at Hilton Head. I was
particularly interested in your presentation of intranasal iontophoresis.
This has considerable potential in autonomic failure, and a whole range of
other autonomic disorders. Please keep me posted on how the work is going
and when there are possibilities of this being tested either in normal man
or in patients.

Otherwise, I trust all remains well and that you continue to flourish.

With best wishes.

Chris Mathias

I visited Dr. Mathias in London at his invitation. After my
explanation, he was very interested and asked if he could be
involved.

Appendix

8 February, 1999

Dr E N Lerner MD PhD
Director
Lerner Medical Technologies Ltd
A J Ernststraat 171
AMSTERDAM
1083 GT
THE NETHERLANDS

DEPARTMENT of VETERINARY CLINICAL STUDIES
Royal (Dick) School of Veterinary Studies

The University of Edinburgh
Easter Bush Veterinary Centre
Easter Bush
Roslin Midlothian
EH25 9RG

Fax 0131 650 6588
Telephone 0131 650 3000
Email: Joe.Mayhew@ed.ac.uk

Dear Dr Lerner

I am sure you will agree that our studies using your electroautoneurography equipment on a horse were very successful.

Undoubtedly there are superficial skin responses that are dramatically altered by various stimuli and most probably obtunded by a low dose of atropine.

The most exciting potential application of this would be in the diagnosis of equine grass sickness. As you know, this is a devastating disease of horses in Northern Europe, particularly in the United Kingdom, and kills hundreds if not thousands of horses a year. Because this is a pan-dysautonomia, it is most reasonable to believe that the superficial skin response and the evoked superficial skin response will be dramatically altered in this disease. This would give us an excellent way of confirming the diagnosis to save considerable suffering and also to improve our treatment of mild cases.

Perhaps even more exciting, though some way off, would be the application of EAG technology to evaluation of potential performance in horses. A lot of work will have to be done before this could be claimed but that is an exciting prospect.

I am certainly very happy to collaborate in the research that will be necessary to continue the application of EAG to veterinary medicine and look forward to a fruitful and exciting venture together.

I am enclosing some reprints of articles on autonomic function and disease in horses.

Yours sincerely

Professor Joe Mayhew
Department of Veterinary Clinical Studies

Enc - Reprints

HEAD OF DEPARTMENT Professor R J W Hellyer-H MA, VetMB PhD MRCVS

KLINIEK VOOR ORTHOPEDIE
FYSISCHE GENEESKUNDE
EN REVALIDATIE

Diensthoofd
Prof. dr. R. VERDONK

FYSISCHE GENEESKUNDE
EN REVALIDATIE
Prof. dr. G. Vanderstraeten
Prof. dr. M. De Muynck
Dr. L. Vanden Bossche
Dr. C. Vander Linden
Dr. S. Rimbaut
Dr. T. Parlevliet

ORTHOPEDIE
Prof. dr. R. Verdonk
Prof. dr. D. Uyttendaele
Prof. dr. P. Burssens
Prof. dr. S. Stuykens
Prof. dr. K. F. Almqvist
Dr. L. De Wilde
Dr. B. Poffyn
Dr. N. Hollevoet
Dr. W. Vanhove
Dr. F. Plasschaert
Dr. E. van Ovost
Dr. E. Janssens
Dr. Ch. Putteman

UNIVERSITEIT GENT

9000 Gent   17. February 2004
De Pintelaan 185/P5

To Lerner Medical Technology Ltd.

ONS KENMERK: **MB/VDS**
(in uw antwoord vermelden a.u.b.)

Concern :   **A letter of recommendation**

The Electroautonomograph (EAG) invented by Dr. E. Lerner was, in both 1998 & 1999, tested by me and my colleagues Dr. P. Mortelé and Dr. Lissens, in the department of physiotherapy and orthopaedics in the University Hospital of Ghent (Belgium).

The Electroautonomograph is a multifunctional device, which can simultaneously register the physiological autonomic systems status of the entire body (several organs and systems, such as cardiovascular, respiration, gastro-intestinal, etc.).

Up to now, no other medical device or method is capable of doing all these registrations at the same time.

The EAG is equipped with a new, highly advanced software package, which allows the results to be demonstrated quantitatively as well as practically.

This device has the possibility to open completely new doors for the diagnosis of nervous system diseases, especially Chronic Fatigue Syndrome, which is not possible without the EAG.

In conclusion EAG, has a wide range of medical applications in neurology, cardiology, physical medicine and rehabilitation, internal medicine. In my opinion, the cost price of each device would be between 10 & 20 thousand euros.

I am very interested in using further this unique device in patients of the department of physical medicine and rehabilitation.

Prof. dr. G. VANDERSTRAETEN.

Afspraken raadplegingen: tel. 09 240 22 51 - fax 09 240 49 75 - orthofysio@ugent.be • Afspraken Kinesitherapie: tel. 09 240 29 84 - fax 09 240 29 88 - Pat.Vandesteene@UGent.be • Operatieplanning: tel. 09 240 26 73 - fax 09 240 52 86- Gay.Demulder@UGent.be • Secretariaat prof. dr. R. Verdonk: tel. 09 240 22 48 - fax 09 240 49 75 - Monique.Depover@UGent.be • Secretariaat prof. dr. G. Vanderstraeten: tel. 09 240 22 34 - fax 09 240 49 75 - Annie.Jons@UGent.be • Secretariaat prof. dr. D. Uyttendaele: tel. 09 240 22 33 - fax 09 240 49 75 - Eddy.Westerlinck@UGent.be • Secretariaat hand- en microchirurgie: tel. 09 240 36 85 - fax 09 240 49 75 - Hoofdcentrum.UZGent@UGent.be • Verantwoordelijke: tel. 09 240 22 64 - fax 09 240 49 75 - Paula.Christiaens@UGent.be • Technisch Verantwoordelijke: tel. 09 240 62 21 - fax 09 240 49 75 - Vicren.Coecke@UGent.be • Wetenschappelijke

Appendix

UT Medical Group, Inc.
Autonomic Function Lab
UT Bowld Hospital
951 Court Avenue, Room 307
Memphis, Tennessee 38163-2222
901-448-5588
FAX: 901-448-6767

Dr. E. Lerner
Lerner Medical Technology LTD. (BV)
1083 GT Amsterdam
The Netherlands

03-15-99

Dear Dr. Lerner

It was a great pleasure for us to meet with you in Memphis, TN. During your short visit and after you demonstrated to us how to use the Electroautonomograph (EAG), we explored different possible applications for it in evaluating patients with autonomic nervous system dysfunctions.

Initially, we hope to start with two groups of patients. In addition to normal healthy control subjects we would like to investigate a group of patients with symptoms of syncope and another with diabetes mellitus of total of 30 to 60 subjects. The test will be done twice for each subject to evaluate the reproducibility. In addition to our instruments which will monitor BP & cutaneous blood flow, ECG, EGG, respiration, Oxygen saturation & skin potentials channels will be used to record heart rate, stomach waves, respiratory rate, Oxygen level & nerve conductivity, respectively. Continues surface skin temperature, blood flow, BP & EEG will be add in the near future. During the test special non-invasive procedures such as deep respiration, forced respiration, head up tilt table test and evoked skin potentials will be performed. All the above parameters (which can be performed by the EAG instrument) have special CTP code numbers which are honored by insurance companies. Average charge for each autonomic function test using EAG is about $ 2000. Major charges due to EKG (code # 95921) ~ $ 230, BP monitoring (95922)~ $ 280, up to 24 hours HRV (93224) ~$ 380, Tilt Table Test (93660) ~ $ 570 , EGG from $ 200-570 (depends on the duration of recording) nerve conduction study (95900) ~$ 500, cardio-vagal changes tested by the autonomic function test (95921) ~ $230 and sudomotor changes in sympathetic skin response (95923) ~ $ 200.

Furthermore, as a director of the Autonomic Function Unit at Univ. of Tennessee, Memphis, TN. and hence, the EAG instrument is a non-invasive device and all the parameters can be performed, recorded and analyzed, I predict a wide different applications for it.

Finally, and since most, if not all, of the parameters have CPT codes, EAG instrument can be accepted without any problem by the FDA.

Again I would like to thank you for giving us the opportunity and looking forward to cooperate with you as soon as possible.

Sincerely,

Dr. Hani Rashed
Director of the Autonomic Function Unit
University of Tennessee.

Faculty of Veterinary Medicine

Utrecht University

To whom it may concern

Department of Large Animal Medicine and Nutrition

address
Yalelaan 16
De Uithof - Utrecht
The Netherlands

mailing address
P.O.Box 80.152
3508 TD Utrecht
The Netherlands
Telefax **31-30 531617

Date   august 30, 1995
our reference
your reference
telephone   **31-30 53 1112
enclosure
Re.

The method and apparatus developed by dr. E. Lerner for examination of the functional condition of the Autonomous Nervous System (ANS) called ElectroAutonomoGraphy (EAG) is of great interest.

The EAG method is based on registration of the skin potential by a special system of amplification and filtering. The equipment was developed and produced by a Dutch firm called CLB Electronics. The apparatus called Electroautonomograph has five channels and works properly.

With the help of dr. Lerner we have succeeded in developing of a method for registration of skin potentials in the horse and in the cow. In our opinion this is the first time that such an experiment has been performed successfully, since we have not found any publication on this topic in the horse or cow.

The experiments performed in our department convinced us that EAG reflects activity of the autonomous nervous system.

The EAG-method could have great perspective in different fields of veterinary science such as:

- diagnosis & prognosis (prediction) for several diseases
- observation of stress-reactions in animals
- possible development of selection system of more productive animals (cows: meat and milk; horse: sports ability)

We are convinced that the EAG-method could have great future in scientific and practical fields.

If there are any questions about the experiments and the data please do not hesitate to contact me at our department.

R.A. van Nieuwstadt, DVM
Specialist Internal Medicine of the Horse

> This appendix shows good results from EAG tests on horses performed by the faculty of the university in Utrecht, Holland.

Appendix

Dr.E.Lerener

From: "Hans Knoop" <hans.knoop@telenet.be>
To: <e.lerner@wxs.nl>
Sent: Friday, August 08, 2008 12:28 PM
Subject: Helmut Hetzel

http://www.helmuthetzel.com/duits/cv1.php

Dear Dr. Lerner,

Above you will find the link to the website of the German correspondent Helmut Hetzel who we are going to meet on August 18. As you see he is a big shot with outlets in both Germany, Switzerland and Luxembourg. Furthermore he is an expert on China and also has access to Chinese media. Last but not least he is the chairman of the foreign press association!

Best regards,

Hans Knoop

Longevity and Health

Dr.E.Lerener

From: "Stutterheim, Martin" <M.Stutterheim@technopartner.nl>
To: "Dr.E.Lerener" <e.lerner@wxs.nl>
Cc: "Ward Mosmuller" <mosmuller@tto.vu.nl>
Sent: Wednesday, October 17, 2007 8:59 AM
Subject: RE: Etdds

Dear Professor Lerner,
If it is okay with you we can meet with mister Mosmuller on wednesday the 31st of
October from 15:00 - 16:00 hours. The meeting will take place in Amsterdam. The adress
is :

De Boelelaan 1085 (W&N, F-545)
1081 HV Amsterdam
The Netherlands

Dr.ir. Ward (E.W.J) Mosmuller MBA
Director Technology Transfer Office VU & VUmc
T +31 (0)20 598 9905
F +31 (0)20 598 9904
M +31 (0)627 050 193
E mosmuller@tto.vu.nl

Please let us know if this okay?
Kind regards,
Martin Stutterheim

-----Oorspronkelijk bericht-----
Van: Dr.E.Lerener [mailto:e.lerner@wxs.nl]
Verzonden: dinsdag 16 oktober 2007 18:44
Aan: Stutterheim, Martin
Onderwerp: Re: Etdds

Dear Mr.Stutterheim,
thank ou for the e-mail. As I told by the telephone, I have
time for a meeteng any day and time.Please as soon as
possible for you and Mr. Mosmuller.
Yours Eduard Lerner.

www.ingramcontent.com/pod-product-compliance
Lightning Source LLC
Chambersburg PA
CBHW031829170526
45157CB00001B/238